*That mystery of mysteries,
the replacement of extinct species by others.*

John F. W. Herschel, astronomer-philosopher,
in a letter to Charles Lyell, geologist, February 20, 1836

MYSTERY OF MYSTERIES

Is Evolution a Social Construction?

MICHAEL RUSE

HARVARD UNIVERSITY PRESS

Cambridge, Massachusetts
London, England
1999

Library of Congress Cataloging-in-Publication Data

Ruse, Michael.
Mystery of mysteries : is evolution a social construction? /
Michael Ruse.
p.   cm.
Includes bibliographical references (p.   ) and index.
ISBN 0-674-46706-X (alk. paper)
1. Evolution (Biology)—Philosophy.   2. Science—Philosophy.
3. Biologists.   4. Scientists.   I. Title.
QH360.5.R874   1999
576.8′01—DC21   98-41969

For

John Beatty

Peter Sloep

Paul Thompson

Preface

This is a book about the nature of science using evolutionary theory as a case study. It follows on the heels of another book written by me, *Monad to Man: The Concept of Progress in Evolutionary Biology,* also published by Harvard University Press—a book which contains similar themes. The difference is that whereas in that book I was using philosophy to try to understand biology, in this book I am using biology to try to understand philosophy. I should say also that *Monad to Man* was directed at a specialized audience: it was long, with detailed references. This book is intended more for a general audience. I have tried not to overburden the reader with background research, a temptation for a scholar somewhat akin to the addict's craving for a cigarette. Although *Mystery of Mysteries* and *Monad to Man* are two independent books and meant to be read that way, if you really want more information or documentation on claims made in the former, I suggest that the best place to start is with the index to the latter.

Monad to Man relied heavily on archival findings as well as on detailed personal interviews. Here, although my understanding of evolutionary theory's history is obviously influenced by this research, I have confined myself to the public printed record. The exception is extensive interviews I conducted with two of today's leading active evolutionists: the English sociobiologist Geoffrey Parker of Liverpool University and the American paleontologist Jack Sepkoski of the University of Chicago. One of the marks of modern professional science is a reluctance of its practitioners to reveal their cultural commitments too publicly, especially in their technical writings. Since these two men are professionals *par excellence*—that is a major reason why I chose them—and since I wanted to know something of their extrascientific lives and beliefs and of the possible connections with their science, lengthy face-to-face discussions were needed. It is the nature of such encounters that Parker and Sepkoski were somewhat at my mercy. Obviously, I could not tell them what I was hoping or dreading to hear before I started the interviews. Hence, here, I want to express not merely thanks for their time and hospitality but the sincere hope that I have not betrayed their trust or our friendship. If they disagree with what I have made of the material, I am sure they will agree that such disagreements are the nature of scholarly discourse.

As always when writing a book, I have incurred debts to many people. The immediate support while I was writing this book came from the Social Sciences and Humanities Research Council (of Canada). It is the final part of a project which began a decade ago when I received a fellowship from the Isaak Walton Killam Memorial Fund. As always, I owe much to my home institution, the University of Guelph, especially my dean, Carole Stewart, and my chair, Brian Calvert. I am much indebted to the following people who read earlier drafts of the manuscript: Barry Allen, Michael Bradie, Jim Brown, David Depew, and Bruce Weber. Denis Lynn read and offered suggestions on the glossary. My secretary, Linda Jenkins, and my research assistant, Alan Belk, did all of those things which I should have done myself but did not. Michael Fisher and Susan Wallace Boehmer of Harvard University Press have been everything that one could hope for in editors: friends, guides, supporters, and critics. Most importantly, Lizzie and the kids were there whenever I looked up from the keyboard.

Contents

Photographs follow page 121; credits on page 258

Mystery of Mysteries

Is Evolution a Social Construction?

"The Einsteinian constant is not a constant, is not a center. It is the very concept of variability—it is, finally, the concept of the game. In other words, it is not the concept of some*thing*—of a center starting from which an observer could master the field—but the very concept of the game."

In mathematical terms, Derrida's observation relates to the invariance of the Einstein field equation $G_{\mu\nu} = 8\pi M T_{\mu\nu}$ under nonlinear space-time diffeomorphisms (self-mappings of the space-time manifold which are infinitely differentiable but not necessarily analytic). The key point is that this invariance group "acts transitively": this means that any space-time point, if it exists at all, can be transformed into any other. In this way the infinite-dimensional invariance group erodes the distinction between observer and observed; the π of Euclid and the G of Newton, formerly thought to be constant and universal, are now perceived in their ineluctable historicity; and the putative observer becomes fatally de-centered, disconnected from any epistemic link to a space-time point that can no longer be defined by geometry alone.

(Sokal 1996a, 221–222; quoting Derrida, above, 1970, 266)

⧓

Science Wars

Very impressive stuff, especially if it comes dripping with footnotes as learned and as obscure as the text. But in the privacy of your own mind—with your guard down intellectually—have you really any idea what the quotations on the opposite page actually mean? "The infinite-dimensional invariance group erodes the distinction between observer and observed"? Although I am not sure that I would have been brave enough to be the first to say so publicly, to me it all reads like pure, unadulterated gobbledy-gook. And I very much hope that it does to you, too, because that is precisely what it is! Nonsense in polysyllables, pretending to be a serious contribution to knowledge.

However, the editors of a major journal, *Social Text,* in the trendy new academic discipline of "cultural studies," did not read it that way. They took the paper seriously and published it. At once, the author, a reputable physicist from New York University, revealed it for the hoax—the pseudo-article—that it is (Sokal 1996b). Whereupon, failing to realize that there are times when the only sensible course of action is to maintain silence, as dignified as you can make it, one of the gurus of cultural studies penned a long and windy and essentially irrelevant opinion piece in *The New York Times,* defending the editors in their silly and (to be frank) slipshod actions (Fish 1996).

Academics love this sort of thing. Even normal people can crack a smile, when seemingly arrogant, pompous, but essentially shallow and

lazy people, who talk in loud, bullying tones on subjects about which
they know absolutely nothing and cloak their nonthoughts in ponderous
imported jargon ("hegemony"—does anybody really know what that
word means?), are shown to be the charlatans that they truly are. And
if they are sufficiently conceited or naive to fight back, then so much the
more fun. For academics, it is time to turn to the keyboard and add to
the controversy. Historians can compare this with great hoaxes of the
past. Philosophers can discuss the ethical implications and whether the
perpetrator, who at once revealed his role, can strictly be considered to
have committed a fraud. And scientists can tell all who will listen that
the affair only shows that English departments, where cultural studies
is usually located, deserve even less funding than they currently get.
Why do they not stick to teaching people how to use the semicolon
properly?

But pull back for a moment. Stop the argument about whether the
physicist-author, Alan Sokal, deserves a medal or censure, or whether
the cultural studies defendant, Stanley Fish, is a man of courageous
integrity or foolhardy insensitivity. Let us put things in context and ask
ourselves why this happened. Why would a serious scientist take time
out to pen a hodgepodge of quasi-fragments about the nature of science,
glued together by the worst excrescences of French philosophy, dolled
up with all of the apparatus of the scholarly article—quotations, footnotes, references—and send it off to a journal not in his field? And why,
why would serious scholars in the humanities—and these people are
very serious—be so eager to receive and accept such a piece that they
would embrace it and legitimize it by putting it in their journal? Why,
above all, would they be so self-confident that they would publish such
a piece without first running it past at least one person who knew
something about physics?

Start with the scientists. In this century, they have had what one
can with modesty describe as a good run for their money, although more
precisely one might describe it as a good run for our money. For various
reasons, this has been the century of science, of great science: relativity
theory, quantum mechanics, the double helix, plate tectonics, and much
more. It has, moreover, been the century of the scientist, as governments, foundations, industry have poured vast sums of money into the

enterprise, producing virtual factories of researchers, technicians, students, administrators, and coordinators, all dedicated to turning out more and more empirical results, more and more theories and hypotheses, in more and more outlets: journals, books, bulletins, conference papers, and various electronic forms.

But now, again for various reasons, the funds are drying up and the prestige is wilting. Part of this is from a general revamping of the global economy, with Western governments and industries having to retrench, to spend more frugally, especially on things without prospects of immediate return—pure scientific research, for instance. Part of this is from the changes in the global power structure, with the collapse of the Soviet communist system and the end of the Cold War. No longer is there the perceived need to spend large sums on defense-related science. Does anyone really think that trips to Mars are needed to save us from the Russians? And part of this decline in the status of science is due to a general change in our culture—an increasing willingness to ask difficult and hostile questions about the sacred icons of society and less willingness to rest content with obfuscating banalities in reply.

This last factor is particularly grating to scientists, for culture does not change in isolation and without reason. In some cases, the enemies of science are obvious and, while dangerous, at least understood and respected for what they are. When biblical literalists try to destroy the teaching of evolution, one may be aroused but one is certainly not surprised. Nor does one feel particularly resentful, or betrayed. After all, what would one expect of an evangelical Christian who accepts the absolute truth of every word of Genesis? But the enemies apparently also reside within, and over this less obvious fact there is resentment, deep resentment. Science is under attack from people of equal standing, often from people inhabiting the same institutions: that is, from scholars in the humanities, from many in the social sciences, and even, in some few cases, from inside the scientific enterprise itself. Always jealous of science and its success, these critics take now the opportunity to attack the empirical investigation of nature and to drag it through a mud of their own making.

How can this be and how can it have come about? The manifesto of the doughty defenders—that which stimulated Sokal to action, not to mention the editors of *Social Text* in their search for science-debunking contributions—appeared a year or two back. Written jointly by a life-sciences administrator and a professional mathematician, Paul R. Gross and Norman Levitt, *Higher Superstition: The Academic Left and Its Quarrels with Science* has an explanation as simple as it is stark. The 1960s was the age of the flower children: sex, drugs, Eastern mysticism, and above all a deep hatred of science, seen to be the essential engine of the military-industrial complex, then engaged in a corrupting and evil, although highly profitable, conflict in Vietnam. Times have moved on, but not the thinking of these children—children no more, but powerful professors and administrators in the humanities and social science faculties of the universities of the West. Now they and their students can give full vent to their opposition to science, an opposition based on prejudice, fear, and, above all, rank ignorance. Searching out allies and molding opinion to their ends, these critics have no limits to their intentions and their arrogance. Little wonder, then, that the editors of *Social Text* seized happily on Sokal's submission—a piece rubbishing the pretensions of modern science and from a scientist himself! Exposing the piece to referees could only lead to criticism, and that is precisely what the editors did not want.

To the outsider, this scenario sketched in *Higher Superstition* sounds like paranoia. Or self-interest. One's suspicions are hardly abated when one learns from Gross and Levitt that a good way to stop the rot would be to put the hiring of new faculty in the humanities in the hands of the nation's scientists. Not only would they be asked to judge the merits of applicants in high-energy physics but also in Restoration comedy. One shudders at the thought. Goodbye *Social Text!* Welcome *Reader's Digest!* Yet, as the saying goes, even paranoids have real enemies. Could it be that these people have a point, that there is indeed a conspiracy or (perhaps with less conscious design) a movement to tear down the status and achievements of science—a conspiracy or movement fueled by ideology, in respects akin to that fingered by Gross and Levitt?

One has to say that precisely this is suggested by the editors of *Social Text*, in their arrogant response to Sokal's hoax. They speak insouciantly of "questioning, as we do, the scientific community's abuses of authority, its priestly organization and lack of accountability to the public" (Robbins and Ross 1996, A28). The chutzpa level is off the scale. Uncontrite, they trust that the kind of critique they level "will help us avoid disastrous scientific irresponsibility in the future."

Stuff like this does not come from nowhere, even from members of English departments. The fact is that people like this are fortified by three or four decades of systematic deconstruction of science, its practitioners, its products, its promoters. Indeed, in respects the literary criticism types are Johnny-come-latelies, noteworthy more for the venom of their attacks than for the originality of their arguments. The materials for critique lie readily at hand. Take some of the real heroes of science. One by one, they have been paraded forth, clad only in their tattered underwear, with signs around their necks, rather like the victims of one of Mao Zedong's purges. In the eyes of their critics, a less creditable, more sleazy bunch would be hard to imagine.

Isaac Newton, for example, the greatest of the great, discoverer of the law of gravitational attraction, author of the *Principia,* mathematical genius. Or so you might think. However, for a start it appears that he was one of the shiftiest data manipulators in the history of physics, the kind of man who sends shudders down the spines of honest researchers and who has congressional critics in spasms of investigative frenzy. He trimmed, cooked, and forged the data until his science was as stylized as a painting by Picasso. And he was so brazen and arrogant about it all: "Not the least part of the *Principia's* persuasiveness was its deliberate pretense to a degree of precision quite beyond its legitimate claim. If the *Principia* established the quantitative pattern of modern science, it equally suggested a less sublime truth—that no one can manipulate the fudge factor quite so effectively as the master mathematician himself" (Westfall 1973, 751–752).

But more than that, we now know that Newton showed far less enthusiasm for sober science and much more for crazy speculation about biblical prophecy and alchemic experimentation. Indeed, not only did he poison himself with foul chemicals, but there is reason to believe that his thinking about gravitational attraction was a mere outgrowth of murky mystical speculations that came from his own strange chemistry: "the alchemical hermaphrodite, sulphur surrounded by its mercury, offered a model to explain the universal property that all bodies possess to act upon each other at a distance" (Westfall 1980, 375). A sadist of the worst kind—his persecution of forgers when he was Master of the Mint turns even hardened stomachs—Newton's only redeeming feature seems to have been that he was homosexual, although even here we find some pretty dicey relationships.

Next comes Charles Darwin, father of evolutionary theory and a real neurotic if ever there was one. Not that we should be surprised since, coming as he did from the comfortable rich upper-middle classes, in proposing a theory of common origins Darwin was threatening the very foundations and stability of the social hierarchy which supported and nourished him. Little wonder that reaction to *On the Origin of Species* was purely political and that those who really reveled in its message were the rank atheists. Fortunately, whatever his failings as a man of science—and they were legion—in other respects Darwin was a man of resource, and so by the time that he published his work on our own species (*The Descent of Man*) the beast had been tamed. Indeed, his vision was essentially the Victorian equivalent of home video (Desmond and Moore 1992, 580):

> The Darwins fitted the picture perfectly. The *Descent* was essentially their story. Natural and sexual selection had made and maimed them. Charles had strutted like "a peacock admiring his tail" courting Emma. Coy and impressionable, she had selected him, admiring his "courage, perseverance, and determined energy" after a voyage around the world. Her "maternal instincts" and feminine intuitions had been the mainstay of their marriage (even if partly a hold-over from "a past and lower state of civilization"). Endowed with wealth, they had a head-start in the struggle—and an "accumulation of capital" was essential if civilized Westerners were to spread and subdue the lower races.

And so on and so forth, right through to the point where evolution's champion had so domesticated the subject that, as a mark of thanks from a grateful nation, he ended up buried in that peculiarly English Valhalla, Westminster Abbey. Appropriately, he and Newton lie there together, for eternity.

Freud hardly deserves a mention. Everyone criticizes him. Even those who come to praise him get in their licks. Having destroyed his personal papers, "Freud actively sought to cultivate the unknown about himself to ensure that he, as intellectual hero, would not be devalued by an overly detailed understanding of his genius" (Sulloway 1979, 7). This is perhaps as well, for it would be a mistake to think that Freud should "be judged against the higher standards of certainty that generally prevail for research and discovery in the physical sciences" (499–500). Little wonder that in right-thinking philosophical circles Freud has become a by-word for how not to do science. The unhealthy relations of a late-nineteenth-century Viennese family have been blown up into a phantasmagoric edifice, totally lacking foundations in the real world.

Albert Einstein might seem more hopeful, but here too one should step with care. In this modern age, all of that stuff about relativity is too obviously cultural to need detailed proof that his work was more an epiphenomenon of his society than a disinterested reflection of objective reality (Gross and Levitt 1994, 46, quoting Ferguson 1990, 238):

> The inner collapse of the bourgeois ego signaled an end to the fixity and systematic structure of the bourgeois cosmos. One privileged point of observation was replaced by a complex interaction of viewpoints. The new relativistic viewpoint was not itself a product of scientific "advances" but was part, rather, of a general social and cultural transformation which expressed itself in a variety of "modern" movements. It was no longer conceivable that nature could be reconstructed as a logical whole. The incompleteness, indeterminacy, and arbitrariness of the subject now reappeared in the natural world. Nature, that is, like personal existence, makes itself known only in fragmented images.

In any case there are serious questions about whether Einstein really did the work himself, or whether the true author was that first wife whom he conveniently dumped, when once he was on the road to fame (Stachel 1995).

So the story goes on. The latest giant of science to receive the debunking treatment is Louis Pasteur, famous for his work on milk and wine and against disease, most notably rabies. I suppose that when you pick up a book with the title *The Private Science of Louis Pasteur* (by Gerald Geison) you already have an idea what it is all about, and you are certainly not disappointed in this case. In the words of one sympathetic reviewer, Pasteur appears "authoritarian, politically reactionary, self-deceiving, overly concerned with priority and credit, ungenerous to his devoted assistants and ruthless with his less privileged and capable adversaries, and overconfident and reckless in putting human patients at risk" (Kohler 1996, 332). And this is just a start. I have but one piece of advice if Pasteur comes around offering relief through some new pharmaceutical product. In the immortal words of Nancy Reagan: "Just say no!" The famous anthrax vaccinations were done with a substance prepared according to a recipe filched from a rival, they had been preceded by very private failures, and they were inflicted on humans before anyone had got round to testing them on animal models.

Enough. No wonder that the science critics are intoxicated with the success of their crusade. Nary an idol of modern science stands erect, untouched. And if you think these men themselves were moral and intellectual cripples, imagine the status of the work they produced. Why should the editors of *Social Text* have bothered to seek out the refereeing expertise of the very group they are determined to destroy? If, as they now say, Sokal's article struck them then as rather ponderous and naive, this is only to be expected and a small price to pay in the campaign to take the battle right to the enemy. Nor, looking at things from the other side, is it any wonder that the scientists are hugging themselves with delight at the abyss into which the critics have tumbled—or chuckle that these critics now sit at the bottom, mud-bedaubed, blaming others. After years of abuse, the tormented giant is striking back.

But there are serious issues here worth discussing. Science is important, important to us all. Whether you think it is more or less than the sum of its parts, including its practitioners, it is today a major part of our culture, in the broadest sense. For all of the attacks, it is a dominant factor in our lives. Moreover, who in this century of war and destruction—evils brought on in large measure by science-fueled technology—could deny that science has its dark side, and that possibly not everything claimed in the name of science is a simple reflection of objective reality? When recently the possibility of life on Mars was announced, was I the only one who smiled cynically as the National Aeronautics and Space Administration at once demanded more funds, and a president felt the urge, in a reelection speech, to praise our great researchers?

The clash over the nature and status of science presents many different aspects. In this book, I am going to isolate what seems to me to be the major issue. This is simply whether science should be considered something different and rather special—something with independent standards which in some way guarantee its truth and importance, and which would merit societal support even if individual scientists are as fallible and untrustworthy as critics maintain. Or is science basically just a product of the same general culture as most everything else, no worse but certainly no better than those who produce it—dangerous precisely inasmuch as it is thought to have some distinct and special status, a mistaken belief which too often comes from the devious self-serving actions of scientists themselves?

The way that many would cast this debate is in traditional philosophical terms. Is science a description or report of an objective reality, a real world which exists independently of humankind and which would be the same even if none of us had ever been? A product where the standing and nature of the producer is totally irrelevant? Or is science better thought of as subjective in some sense, a creation rather than a discovery, a thing rising from and yet bound by culture, an artifact of humankind that would not necessarily be the same in different places and at different times?

Being a philosopher myself, I am much aware that we philosophers have a bad reputation when it comes to debates like this. Supposedly we delight in drawing distinctions which are so fine-grained that eventually everything disappears into language and we end up triumphantly proclaiming that no real debate existed in the first place. Let me therefore assure you at once that I think there are real issues here and real differences. I am enough of a traditionalist, however, to think that different matters are being conflated, and that we shall need to tease them apart to see what can and what cannot be answered, and by what means. But for here and now, I am quite happy to roll everything together into one fundamental divide, between science—with special

standards or without, beyond culture or within, objective or subjective—
and its critics. I am aware that one might be a critic of scientists'
pretensions to objectivity and still think that the products of science are
a jolly good thing, although as a matter of fact many if not most
subjectivists are also critics in the sense of not much caring for the
practices and products of modern science.

However you frame it, this is the debate of frontal concern on both
sides. The attack on character does hurt. Science is right up there with
sport in making a cult of the heroes of the past. When the Nobel
laureate Max Perutz (1995) wrote a violent and bitter review of *The
Private Science of Louis Pasteur,* he spoke in rebuttal of the French
scientist as "courageous, compassionate, and honest." Yet much more
than character is at issue, for even if science is a creation, one could
argue that logically the character of Pasteur is irrelevant to the truth and
importance of his theories. A total rotter might produce something of
transcendent worth. Think of Wagner and his operas. But both Perutz
and the author of the book he was reviewing know that this debate goes
beyond logic. The very heart of science is at stake.

For Perutz, the true worth of Pasteur lies in the truth and utility of
his theories: "his scientific achievements, which have much reduced
human suffering, make him one of the greatest benefactors of mankind"
(58). And this was no pure chance, for we are talking here of discovering
reality rather than creating an artifact. Perutz indeed headed his review
with a quotation from the physicist Max Planck: "There is a real world
independent of our senses; the laws of nature were not invented by man,
but forced upon him by that natural world. They are the expression of
a rational world order." You cannot be less ambiguous than that.

Though you can equal it, which is what we find on the other side.
Even those who believe in a real world—by no means a general assumption—argue for the deep human-based relativity of science. Consider
Stanley Fish's response to Sokal's hoax, a response in which he compares baseball to science (1996, 23):

> On the baseball side, the social construction of the game assumes
> and depends on a set of established scientific facts. That is why the
> pitcher's mound is not 400 feet from the plate. Both the shape in

which we have the game and the shapes in which we couldn't have it are strongly related to the world's properties.

On the science side, although scientists don't take formal votes to decide what facts will be considered credible, neither do they present their compelling accounts to nature and receive from her an immediate and legible verdict. Rather they hazard hypotheses that are then tested by other workers in the field in the context of evidentiary rules, which may themselves be altered in the process. Verdicts are then given by publications and research centers whose judgments and monies will determine the way the game goes for a while.

Both science and baseball then are mixtures of adventuresome inventiveness and reliance on established norms and mechanisms of validation, and the facts yielded by both will be social constructions and be real.

In fact, with his snide comment about the "monies" of the research centers determining the fate of science, I am not sure how far Fish would commit himself to the ultimate reality of nature. Today, rich and powerful men control most of our newspapers, and we all know what that has entailed in matters of truth and reality. But leave the personal and turn to the important general issues. You might think that this debate—science: objective or subjective?—is one where a little common sense, perhaps a dab of logical clarity, would put things right. Strip away the emotions and look at things with cold reason, and surely the right answers will emerge. Perhaps all we need is for scientists to tell us honestly what they think and feel and find, and then the truth will prevail. After all, it is the scientists who are at the coal face, as it were. If they stop blustering and snarling and presuming, perhaps we can all move forward. Some scientists may deceive some of the time, but not all scientists all of the time.

Would that things were quite this easy! I am not going to be a total cynic and suggest that everything a scientist says is necessarily corrupt, self-serving, or unreliable. But a charge akin to this is certainly involved, for it is precisely the claim of the critics that—even when they do not intend to deceive—scientists (like the rest of us) are pawns of their culture, their interests, their patrons, and much more. Hence, whatever the sincerity, science simply cannot be taken at face value. The successful scientist is precisely the person who is, at some level, morally and culturally insensitive, if not dead. If this does not come about through self-selection, then it is a function of the training, especially those long years spent in graduate and postdoctoral study. The sanctification of the great scientists of the past is part of the indoctrination—the ideology that science stands beyond ideology.

How is one to answer the difficult questions raised by these charges? If you take seriously Whitehead's aphorism, that the whole of philosophy is a series of footnotes to Plato, you might think that we should start back in ancient Greece. Actually, in this instance it would not be such a silly idea. In epistemology—the field of philosophy which studies how we know what (we think) we know—Plato wrestled precisely with the problem of appearance and reality: How is one to reconcile the mind-independent reality of existence with the obviously mind-dependent nature of so much of what we believe and think and use in everyday life? Plato argued that only the ideal Forms are truly real and that we in this world make do with mere reflections or shadows.

Tempting though it is to spend time lingering in Plato's world, it is necessary to move quickly to the problem at hand. I shall therefore turn to look at the ideas and influences of the two major figures in the post–World War II period who have commented philosophically on the nature of science. These two men, the Austrian-born English philosopher Karl Popper and the American physicist-turned-historian-turned-philosopher Thomas Kuhn, both recently dead, pointed us toward the twin ends of the spectrum and have had immense influence. By exploring their ideas and influence, we shall be able to understand more fully the debate today about the nature of science—the points made in favor of the two perspectives and the points against them. Thus informed, we shall be in a position to see how we might start to move forward in a constructive manner.

1

KARL POPPER AND THOMAS KUHN

Two Theories of Science

The irony is that, in their ways, both Popper and Kuhn were outsiders. Not in obvious respects, of course. Popper was a professor at the London School of Economics, a knight, a Companion of Honour, a Fellow of the Royal Society and of the British Academy. Kuhn was a professor (first at Berkeley, then at Princeton, and finally at MIT), one of the best-selling academic authors of all time, a member of the National Academy of Sciences, and, toward the end of his life, president of the (American) Philosophy of Science Association. But neither Popper nor Kuhn fit into a conventional mold or had the unqualified support and admiration of their peers. The things that were said about Popper were downright nasty, and the name of Kuhn could curl the disdainful lip of any right-thinking graduate student. He may have been president of the philosophers of science, but it was long after lesser men had held the post.

There were various reasons for this, not the least of which were personal. Popper could be deeply charming, and as a speaker he was truly charismatic. He could hold his audience enthralled for an hour, to thunderous applause. But he was a dreadfully insecure man, seeing plots when they existed and when they did not, surrounding himself with sycophants who shoveled on the flattery, and exhibiting a pathetic touchiness about criticism. It was hard not to be irritated by a man like that, and when people put pen to paper, they tended to be scathing—especially those who had taken the precaution of remaining ignorant about what he had actually written (Callebaut 1994).

Kuhn was different, both as a man and a thinker. He garnered genuine respect and affection, and as a historian of science, where he first made his mark and at which he worked all of his life, he was always recognized as a master. Disagreements were never hostile or bitter. Yet there was always the unspoken, sometimes spoken, sense that as a philosopher Kuhn was somewhat shallow. He had not penetrated the depths that others, particularly those others adept at the use of formal techniques, had fathomed.

The real problem for both men was that they were great communicators. They had the ability to write clear, hypnotically compulsive prose, using felicitous terms and phrases which entered the common speech, reaching over or around their fellow professionals to people outside their field, be they Nobel laureates, students, or journalists. Nothing more irritates normal professionals, trained as they are to produce material that will molder in the stacks unread for generations to come, than someone who can do just this. Yet I would argue that men like Popper and Kuhn really do have genius, and this genius lies in their simplicity. They have a vision, in this case a vision of science, which captures something missed by others, and because there is something valuable or worthwhile in this vision, it can appeal to regular folk, sparking a flame of recognition and enthusiasm. Their laurels were well deserved.

Karl Raimund Popper (1902–1994)

The Austrian-born Karl Popper was a talented student in the years after the First World War, performing well in the examinations at Vienna University and completing his doctoral degree with a thesis on Hume's analysis of causation (Popper 1974). These were tumultuous times, and Popper's studies were interspersed with other activities, including work with difficult children and apprenticeship to a cabinetmaker. Realizing that his talents lay with his mind rather than his hands, Popper became a schoolteacher, and it was then, at the beginning of the fourth decade of this century, that Popper wrote his masterwork, *The Logic of Scientific Discovery*, a book which did not appear in English until 1959.

Forced to leave home because of the rise of the Nazis (although Popper's parents were converted Lutherans, they were Jewish by origin), Popper spent the war years as a lecturer in New Zealand. Then, his fame growing because of his social writings, especially an attack on authoritarian philosophies which he penned in *The Open Society and Its Enemies* (1945), Popper was called to the London School of Economics, where he spent the rest of his career. That career, to be candid, was less about innovative new moves and more about refining and defending ideas and insights achieved previously. At this time he became something of the darling of the scientific community, in a way its official philosopher, although whether this was because he accurately described the process of science or because he provided an ideology that science's practitioners found comforting has no ready answer. Indeed, ultimately, it is the question I address in this book.

Austria at the beginning of the 1930s was the home of the famous Vienna Circle and of its philosophy of logical positivism (Achinstein and Barker 1969). Popper was never on the innermost ring—it has been suggested that some of his insecurities stemmed from this fact—and he spent much of his life indignantly repudiating connections between his thinking and theirs. Untangling the links, real and imaginary, is the job of the historian or the disciple. What one can say is that everyone who was touched by the Circle shared the conviction that science is a good thing and is the exemplar of the very best kind—perhaps, with logic and mathematics, the only kind—of knowledge that we have. Notoriously, the logical positivists declared that everything else is meaningless—a declaration which led to all kinds of implausible contortions when it came to things like ethics.

Popper was never so dismissive. Having immersed himself deeply in Kant's philosophy, he was always more sympathetic to metaphysics. But he, like the positivists, ever cherished science as the beacon, the Platonic ideal, of human inquiry. One does not have to be too much of a psychologist to realize that for all of these men, living as they had done and continued to do in a highly unstable and often dangerous society, the status that was given to science satisfied deeply personal emotional needs. Here was something that decent people could hold as sacred.

In order to understand Popper's philosophy—and from now on I

shall present the mature position, ignoring developments, debts, and
other sidelights—it is best to go straight to its heart and to the fact that
Popper was determinedly and absolutely a *realist*. This is one of those
terms which can mean anything to anybody, as can its opposite, *idealist*,
so let me say that for Popper at a very minimum realism meant that the
world exists and that it exists in some way independent of us. Trees
really do fall in forests when no one is around to hear them. They would
do so even if no humans had ever existed, past, present, or future. The
aim of science is to map this reality; Popper speaks often of science as a
"net" which tries to capture reality in its folds.

Speaking formally, the main instrument of the scientist is the
theory, which for Popper took on the traditional form of a "hypothetico-
deductive system." The world is seen as governed by regularity, and
theories try to map these regularities through high-powered general
claims, called hypotheses. These hypotheses serve as axioms in a deductive scheme from which lower-level theorems or laws can be derived,
and it is through these latter empirically based claims that science lays
itself open to check and test by physical nature.

Popper saw science as a dynamic process. We come out of our past,
we go into our future. One never starts from scratch, with bare sensa-
tion. Always, one starts with information, presuppositions, ideas, prejudices. All observation is theory-laden. The scientific process gets going
when this gathering of material throws up anomalies, *problems*, which
call for explanation. Then one proposes a tentative solution, a hypothesis, which one checks and tests. If it works, then so far so good. But
eventually, there is always the possibility of something breaking down,
a fresh problem being generated, and the scientific process starting all
over again. Popper held to a correspondence theory of truth—truth
consists in getting your ideas to match reality exactly—and thought that
such truth is in principle possible. But one can never be sure that one
has arrived at the truth. The best one can hope for is an ever-more
precise approach to an understanding of reality: ever closer, but never
absolutely certain that one is there.

Popper (1974) therefore saw science as a kind of Darwinian process,
where ideas compete in the marketplace and where, after rigorous
selection, the best ones survive—best for a day always, and with new

rivals ever on the horizon. This made him what the social psychologist Donald Campbell (1974) labeled an "evolutionary epistemologist."

Popper laid things out in the following fashion:

$$P_1 \rightarrow TT \rightarrow EE \rightarrow P_2$$

A problem, P_1, inspires a tentative theory, *TT*, which is then subject to test and to error elimination, *EE*, giving rise in turn to another problem, P_2. Popper was very much aware that often one has several problems and theories all mixed up together and that the actual process of science is a lot messier than this simple sequence implies.

With this schema we are able to see clearly the most famous aspect of the Popperian philosophy, that for which he is best known. The positive or successful test of a hypothesis can never be final. However well-confirmed (Popper liked the term "corroborated") a hypothesis may be, it can always be overthrown in the future. The possibility of negative evidence is ever looming. But this is no cause for despair, for this potential for refutation is the mark of genuine science: this is the "criterion of demarcation" which separates genuine science from the rest. Real science is *falsifiable*. Not that it is false! That would never do. Rather, it lays itself open to check and, should nature tell against it, rejection. All else is metaphysics or pseudo-science or worse. Metaphysics as such is not wrong; metaphysics pretending to be science is wrong.

One technical problem—the source of the most common criticism of Popper's philosophy—is that posed by ad hoc hypotheses. You can always protect a favored belief by invoking a protective supplementary claim: the instruments were not working properly, there are unknown distorting factors, the calculations need doing in a different way, or whatever. Popper was aware of this problem, and, without judging whether he ever supplied a satisfactory response, one can say simply that he agreed not only that one can but that sometimes one should invoke such hypotheses. Dogmatism can be a virtue. But ultimately, such hypotheses are allowable only if in some way they increase the extent to which a theory lays itself open to check. In their way, protective hypotheses must be part of a forward-looking, ongoing research program, rather than something concerned only with saving what you have.

So what, finally, do we end with? Science, the best kind of science, yields *objective* knowledge. It is what Popper (1972), in a felicitous phrase, referred to as "knowledge without a knower" (109) in the sense that it is independent of the individual scientist who produces it. Male science, or European science, or Jewish science were simply impossibilities for Popper. Science, which Popper assigned to the domain of disinterested ideas (what he called "World 3"), is to be distinguished from mere objects ("World 1") or subjective belief ("World 2"). For this reason Popper stressed strongly the division between what has been called "the context of discovery" and "the context of justification." The English title of his major work, *The Logic of Scientific Discovery*, was misleading, for he did not think there was a *logic* of discovery at all. This area of scientific labor is inspiration or guesswork—brilliant inspiration or guesswork. The logic comes into play when we are systematizing and testing our hypotheses. To think otherwise is to commit what is called the error of psychologism: "The question how it happens that a new idea occurs to a man—whether it is a musical theme, a dramatic conflict, or a scientific theory—may be of great interest to empirical psychology; but it is irrelevant to the logical analysis of scientific knowledge" (Popper 1959, 31).

Enough of Popper's philosophy. As I have said, I am not particularly concerned about its originality, or the extent to which it overlapped with that of others, or whether Popper successfully defended himself against criticisms. My aim is to present what one might fairly call the "objectivist" view of science in its clearest and most popular form. This I take to be the standard view, modernism if you like, certainly the legacy of the Enlightenment, against which the critics rail. And to begin the attack, let us turn to Popper's great rival.

Thomas Samuel Kuhn (1922–1996)

Kuhn was American. Educated at Harvard, he was finishing his PhD in physics when he found himself teaching a course on science to non-science students. Trying to make the subject meaningful through links with the past, he was hooked at once; the teacher became student again, and he retooled as a historian of science. But Kuhn's interests were always more broadly directed at the general nature of science—spurred in part by three years as a Junior Fellow at Harvard, which led to the weekly stimulus of W. V. O. Quine, the doyen of American philosophers. A decade later, in 1962, this produced *The Structure of Scientific Revolutions.* Curiously, the work appeared first as one volume in a series, the International Encyclopedia of Unified Sciences, the definitive organ of the logical positivists! Because the series set very strict limits on the length of contributions, Kuhn was forced to write in a style far more direct and forceful than one usually finds in works devoted to the analysis of science. He had to make his points and move on, rather than killing them with the thousand cuts of the footnote.

The key concept in Kuhn's sparkling volume is that of a paradigm. As friend and foe have pointed out many times, Kuhn means different things by this term at different points in the book. But the chief sense is that of a work or body of work which captures the scientific imagination—which commands allegiance from a group of workers and provides tasks for them to undertake (Kuhn 1962, 10):

> Aristotle's *Physica,* Ptolemy's *Almagest,* Newton's *Principia* and *Opticks,* Franklin's *Electricity,* Lavoisier's *Chemistry,* and Lyell's *Geology*—these and many other works served for a time implicitly to define the legitimated problems and methods of a research field for succeeding generations of practitioners. They were able to do so because they shared two essential characteristics. Their achievement was sufficiently unprecedented to attract an enduring group of adherents away from competing modes of scientific activity. Simultaneously, it was sufficiently open-ended to leave all sorts of problems for the redefined group of practitioners to resolve.
>
> Achievements that share these two characteristics I shall henceforth refer to as "paradigms," a term that relates closely to "normal science." By choosing it, I mean to suggest that some accepted examples of actual scientific practice—examples which include law, theory, application, and instrumentation together—provide models from which spring particular coherent traditions of scientific research.

Science which does not have a paradigm is properly considered immature. It is in a "preparadigmatic state." Once a paradigm has been found, then people can set to work—for Kuhn, science is as deeply a dynamic process as it is for Popper. The paradigm sets the rules, it gives the challenges, it marks out the limits, and much more: it gives rise to "normal science." This is the science that most scientists do all of their lives all of the time. It is science where—and this is absolutely crucial for Kuhn—the paradigm is taken as a given, not to be challenged. It is, in a sense, derivative or clean-up work. "No part of the aim of normal science is to call forth new sorts of phenomena; indeed those that will not fit the box are often not seen at all. Nor do scientists normally aim to invent new theories, and they are often intolerant of those invented by others" (24). When you are working within the paradigm, there is nothing outside the picture.

Problems in science therefore are not problems in the sense of things calling for answers which may or may not be solvable—the Israeli/Palestinian problem, for instance. They are rather puzzles, in the sense that a scientist doing normal science assumes as part of the game that an answer can be found—as in a crossword, for instance. Failure to find such an answer reflects badly on the scientist, not on the paradigm. In compensation, however, the paradigm does declare certain issues off limits. Some problems, "including many that had previously been standard, are rejected as metaphysical, as the concern of another discipline, or sometimes as just too problematic to be worth the time" (37). With supposed consequences like this, many critics, particularly the Popperians, accused Kuhn of reducing science to the humdrum, the boring. He defended himself vigorously against this attack; but, in any case, he argued that it is precisely because science puts on these blinkers that it is so successful. There is a single-mindedness about forward-moving "normal" science.

Now for the revolutions. Every so often, things in science start to come apart. The puzzles that need solving seem to multiply beyond the normal or the expected. Anomalies abound. The paradigm starts to come unstuck. This does not mean that one abandons it. To do so would mean the end of science. If anything, one works more frenetically to shore things up. Then, however, if one is lucky, someone—usually someone young or new to the field (that is, knowing the issues but without great emotional attachment to the old paradigm)—puts forward a new paradigm. It solves or avoids the difficulties of the old paradigm, at the same time that it offers the prospect of much new work in its own right. The community's allegiance switches to this newcomer, and in just a short while normal science resumes again. Moreover, in Orwellian fashion, the textbooks—a key item in the culture of science—are rewritten, and it now appears that the new paradigm was, all along, the logical outcome of the progress of science. One might say that revolutionary science is sanitized into normal science.

In these revolutions, invisible or not, we get the most controversial aspects of Kuhn's vision of science. For him, the paradigms set the rules and reasons for science. Across paradigms no reasonable arguments are possible. The two sides are "incommensurable," to use Kuhn's term. At best one can appeal to such factors as the simplicity or elegance or fruitfulness or metaphysical sympathy of one's chosen position. Scientific revolutions are therefore, if not irrational, at the least arational—outside reason (Kuhn 1962, 94):

> Like the choice between competing political institutions, that between competing paradigms proves to be a choice between incompatible modes of community life. Because it has that character, the choice is not and cannot be determined merely by the evaluative procedures characteristic of normal science, for these depend in part upon a particular paradigm, and that paradigm is at issue. When paradigms enter, as they must, into a debate about paradigm choice, their role is necessarily circular. Each group uses its own paradigm to argue in that paradigm's defense.

Kuhn spends some time in his brief essay showing how seemingly compatible paradigms are no such thing. Shared terms, such as the term "mass" in Newton and Einstein, in fact mean very different things. You cannot compare them, speaking of one as true and the other false, in some absolute sense.

What does this all mean for realism of a Popperian kind? Kuhn has little truck with it. For him, in some very significant sense the world itself changes across paradigms. Taking the notion of theory-ladenness to the extreme—Kuhn is very fond of gestalt-type examples, where once you saw a rabbit and now you see a duck—he argues that the world exists only to the extent that it is seen through a paradigm. "Outside the laboratory everyday affairs usually continue as before. Nevertheless, paradigm changes do cause scientists to see the world of their research-engagement differently. Insofar as their only recourse to that world is through what they see and do, we may want to say that after a revolution scientists are responding to a different world" (111). Paradigms structure observation and hence define reality. Lavoisier saw oxygen where Priestly saw dephlogisticated air. "At the very least, as a result of discovering oxygen, Lavoisier saw nature differently. And in the absence of recourse to that hypothetical fixed nature that he 'saw differently,' the principle of economy will urge us to say that after discovering oxygen Lavoisier worked in a different world" (118).

Is this all just a question of saying that the two sides interpret things in different ways? No, replies Kuhn. In some fundamental way, things (not just ideas or perceptions) really had changed (121–122):

> What occurs during a scientific revolution is not fully reducible to a reinterpretation of individual and stable data. In the first place, the data are not unequivocally stable. A pendulum is not a falling stone, nor is oxygen dephlogisticated air. Consequently, the data that scientists collect from these diverse objects are, as we shall shortly see, themselves different. More important, the process by which either the individual or the community makes the transition from constrained fall to the pendulum or from dephlogisticated air to oxygen is not one that resembles interpretation. How could it do so in the absence of fixed data for the scientist to interpret?

None of this is to imply that the world is unreal, in the sense of being ghostly or dreamlike: "the world does not change with a change of paradigm" (121). But it is to imply much more than the Popperian would allow: "the scientist afterward works in a different world" (121). For this reason Kuhn is inclined to downplay or deny the hallowed distinction between the context of discovery and the context of justification. At some level, one simply cannot distinguish the person from the science. There is no knowledge without a knower, and how you come to accept a new paradigm might be crucially dependent upon your history.

At the very least, acceptance or rejection of a paradigm is going to be dependent not on reality (whatever that might mean) but on what rivals, real or potential, already exist. For this reason, when it comes to talking about the course of science, although Kuhn wants to talk of "progress," it is not progress in the sense of getting an ever-more faithful rendering of (more true picture of) an objective reality. Rather—and note the paradox that he, like Popper, appeals to Darwin—it is the progress of evolution, getting ever-more sophisticated and complex but without any ultimate end point. "The entire process may have occurred, as we now suppose biological evolution did, without benefit of a set goal, a permanent fixed scientific truth, of which each stage in the development of scientific knowledge is a better exemplar" (172–173).

I would not say that Kuhn's philosophy does not overlap with Popper, other than in its appeals to the same evolutionary authority. But there are major differences. Most crucially, for Kuhn "knowledge without a knower" is a contradiction in terms. Reality is defined by the paradigm, and this brings in the scientist. In this sense, knowledge is crucially subjective. Kuhn never claims that "anything goes" or that you can believe just what you want; but for him knowledge is relative—relative to the paradigm, that is. We do not have access to some independent reality against which we can check the contents of paradigms.

Hence, falsifiability is a vain hope. The paradigm creates our reality, and to give it up when faced with problems or anomalies is to give up doing science—unless you have an alternative paradigm. Note "what scientists never do when confronted by even severe and prolonged anomalies. Though they may begin to lose faith and then to consider alternatives, they do not renounce the paradigm that has led them into crisis" (77). Rather, "once it has achieved the status of paradigm, a scientific theory is declared invalid only if an alternate candidate is available to take its place" (77). But this alternative likewise makes its own reality. Hence, what Popper sees as integrity—giving up cherished hypotheses at the demand of contrary evidence—Kuhn sees as stupidity or faintheartedness. What Kuhn sees as good sense—refusing to give way when things go wrong—Popper sees as dogmatism and writing oneself out of science.

Social Constructivism

Of course the established philosophers did not care for what Kuhn had to say, and they had little difficulty in showing that, conceptually, his position was hopeless (Shapere 1964; Scheffler 1967). The notion of incommensurability was a favorite target. But this kind of response was to be expected, and no one whose mind was not already closed took it too personally or seriously. One might have to regurgitate it on one's comprehensive examinations, but that was about the level it was good for. More interesting was the fact that when Kuhn's fellow historians turned to his ideas, generally they were critical, aided by the curious fact that Kuhn himself, when writing as a historian, did not use his own categories. You search in vain for a paradigm-based analysis in his classic *The Copernican Revolution*. Indeed, the one group whose members were enthusiastic for both the language and the ideas was the geologists, who at that time were going through the biggest revolution in their science's history. To a person, they affirmed that plate tectonics is a paradigm *par excellence* (Ruse 1989). After years in the wilderness, they just loved their normal science! This made them the envy of the social scientists, who were stuck fatally in a preparadigmatic phase.

But Kuhn did have an effect—an incredible effect—on people like myself, who were young at the time and just beginning our careers as professional philosophers and historians of science. Given that Kuhn was then a professor at Berkeley, I suppose there was some connection between his ideas and the *Geist* of the age. But while I would not want to deny that any of these things—the Beatles, transcendental meditation, and lots and lots of sex—hit the Guelph campus where I was a very junior faculty member, there were other, more significant if less glamorous and global, factors at work.

Most importantly, the mid-1960s was just the period when the history of science was being professionalized, with teachers, students, and standards. Hitherto, the field had been mainly the province of retired scientists (often of great stature), who would switch from the burdens of administration to the writing of hagiographies of the great men in their specialty. They worked from the printed material on the shelves of their libraries, considering only the scientific ideas in themselves and showing a straight line of improvement or progress up to the present, in which the hero played a crucial role.

Trained historians would have none of this. They went to the archives: to the notebooks, the unpublished letters, to the accounts, the exam papers, the scraps of autobiography. They dismissed the "internalism" of the scientists, arguing that one must be an "externalist" in looking at causes and reasons beyond the strict science. (We can discern a certain amount of self-serving here, for the new historians were often of limited, or zero, scientific background.) And above all they deplored what they called "Whiggishness": the philosophy of history which sees progress from a primitive past to a sophisticated and sound present. Ideas, claimed these new professionals, must be considered in the context of their time, in their own right (Young 1985).

A Popperian philosophy was anathema to people such as this, for all that some of Popper's groupies tried to provide history which fit his pattern. The argument against psychologism (that the context of discovery has no bearing on the context of justification) meant that most of the work—and all of the nice juicy findings—of the externalist historian were ruled out right there, before one even started. The concentration on ideas (internal history) meant that external factors were considered irrelevant; and when scientists got ensnared in violent and distasteful (well, let us not be hypocrites, compulsively viewable) disputes, which they frequently do, this must be considered an aberration. And if *Logic of Scientific Discovery* is not a paean to progressionism and a foundation for Whiggism, then nothing is.

Kuhn, on the other hand, welcomed all that the historian wanted to do—and more. After all, he was one of them! Forget the details, he legitimated the approach. Even if one had trained as a philosopher rather than a historian, one just knew that there was pay dirt here. Philosophers at that time tended to restrict discussion to overly simple or artificial examples: "All swans are white." "All of the screws in my car are rusty." "All emeralds are green before time *t* and blue after time *t*." That sort of thing. Here now were people dealing with "science red in tooth and claw," to mention the overblown subtitle of one of my own Kuhn-inspired earlier books (Ruse 1979). Young philosophers, like young historians, wanted to get on the bandwagon, and so we did. If not the letter, the spirit and the style of *The Structure of Scientific Revolutions* was absorbed.

For all of their empathy, students of science soon realized that there were places where one had to go beyond the word of *The Structure of Scientific Revolutions*. Most particularly, beyond the internalist/externalist dichotomy. Although Kuhn pointed the way forward, in writing about the theory of science he was really quite conservatively internalist about the causes of change both in normal science and at revolutionary times. For the former, it was evidence and puzzle solving; for the latter, it was things like aesthetic appeal and predictive fertility. But as others soon noted, indeed as Kuhn himself noted when working as a historian, external factors seem frequently to be very significant.

One thing that historians delighted in showing is that, contrary to the usually held tale of science and religion being always opposed—the "warfare" metaphor beloved of nineteenth-century rationalists and their twentieth-century counterparts—religion and theologically inclined philosophy have frequently been very significant factors in the forward movement of science. Even the Copernican Revolution, the unchristian implications of which supposedly forced the ancient Galileo to recant his errors, was successful in major part precisely because its enthusiasts found in heliocentrism support for their religious and philosophical prejudices. Every early Copernican was an ardent Pythagorean/Platonist, thinking that a sun-centered universe is spiritually far superior to a universe that puts the earth at the center (Kuhn 1957).

Marxist historians had a field day here in pursuit of the external, showing the extent to which scientific claims were based in social attitudes and beliefs. As did the sociologists, for whom the relativism, the arationality, the subjectivity of the Kuhnian philosophy became virtual orthodoxy—melded, that is, with a strong dose of sociological determinism. "Reality seems capable of sustaining more than one account given of it, depending upon the goals of those who engage with it; and . . . those goals included considerations in the wider society such as the redistribution of rights and resources among social classes" (Shapin 1982, 194).

The writer of these words, Steven Shapin, was one of a group of people in Edinburgh who endorsed what they labeled the "strong programme" for the history and sociology of science—arguing that the supposed truth or falsity of science is not relevant to the historian's task. Nor is the hallowed philosophical distinction between reasons and causes—the distinction between a justification for a belief (such as is provided in a mathematical proof) and an account of how one comes to hold a belief (such as the guilt which leads Macbeth to see the dagger) (Bloor 1976). In a somewhat self-confirming fashion, all of this went along with a new interest in fringe or pseudo-sciences. After all, they were as good as anything else, weren't they? Phrenology became a particularly favored subject of study (Cooter 1984). To privilege (say) astronomy above it would be to accept the very philosophy of internalism and progressionist triumph that was being attacked.

Even when orthodox science was studied, the idealism of the strong program took firm hold. Turning from the past to the present, some sociologists went right into the scientific laboratory, studying scientists as an anthropologist would study some primitive tribe. Naturally, the subjectivities and personalities and controversies and trivialities of everyday life came to the fore, and these were taken to be significant and all-embracing facets of not only the production but also the products of science.

One who took this path was the French sociologist Bruno Latour, who was explicit and forceful in his historicism, his nonrealism, his social constructivism. With a colleague, he wrote as follows (Latour and Woolgar 1979, 128):

> One important feature of our discussion so far is worth noting at this point. We have attempted to avoid using terms which would change the nature of the issues under discussion. Thus, in emphasizing the process whereby substances are *constructed*, we have tried to avoid descriptions of the bioassays which take as unproblematic relationships between signs and things signified. Despite the fact that our scientists held the belief that the inscriptions could be representations or indicators of some entity with an independent existence "out there," we have argued that such entities were constituted solely through the use of these inscriptions. It is not simply that differences between curves indicate the presence of a substance; rather the substance is identical with perceived differences between curves.

So much for Popperian reality: "Our point is that 'out-there-ness' is the *consequence* of scientific work rather than its *cause*" (180–182).

Toward a Resolution

There were and are many like-minded critics of the traditional realist position. Highly influential has been the French philosopher-historian Michel Foucault (1970), whose notion of an "episteme"—a kind of collective unconscious of the age—bore significant similarities to the paradigm of Kuhn. But here I have little interest in providing a cata-logue of such people. It is enough to note that, beyond the nonstop torrent from historians demonstrating the culture-impregnated nature of science, the range of people who have boarded the relativistic, sub-jectivist, social-constructivist bandwagon is wide. Particularly vociferous have been the students of rhetoric, who argue that science is as much a reflection of society as any Victorian potboiler and that one can draw no true distinction between so-called Great Science, like Watson and Crick on the double helix, and the most trivial or indifferent or border-line productions (Gross 1990). Suffice it to say, these people, like many others in the nonscientific community, join with the historians in reject-ing the vision of a Popper, or indeed the vision of most progressive thinkers since the Enlightenment.

In the face of such criticism, it is small wonder that scientists, particularly those who have considered themselves to be sophisticated people of the left and who much resent the ascribed taint of rightist ideology, are striking back and applauding those who land telling blows. More than one ecologically sensitive cup of coffee has been raised in toast of Sokal. More than one laboratory head has recom-mended Gross and Levitt's *Higher Superstition* to his underlings. More than one researcher has regretted the day that cultural studies got a toehold on campus. But can one do more than this? Can one argue reasonably against the attack on science? Can one see if indeed the critique of the traditional philosophical picture of science has points of merit and yet find some way of preserving that which is good and worthwhile about science? This is my hope in this book: a positive answer to this question.

But how is it to be achieved? How is one even to start to find an answer, let alone a satisfactory one? I have little sympathy for those who would, ostrichlike, ignore or deny the work of the subjectivists. Whatever you might think of the scholarship of particular individuals, overall the material that has been uncovered about the history and sociology of science—the ideas and the actors—is simply stunning. Moreover, many of the causes that are espoused have, at the very least, an initial plausibility. Isaac Newton *did* fiddle his figures, he *did* spend a huge amount of time deeply engaged in alchemy, and it is at least arguable that his beliefs about action at a distance were sparked by his esoteric inquiries. It is by no means obvious that all such investigations, considerations, and historical evidence are irrelevant to Newton's work or its reception.

Even if in the end one wants to make of it something different—perhaps something less, perhaps nothing at all—from what the historians and sociologists and others would claim, one has got to look at material like this. "History, if viewed as repository for more than anecdote or chronology, could produce a decisive transformation in the image of science by which we are now possessed" (Kuhn 1962, 1). Even if one is not to end with a Kuhnian philosophy, one has to look at real science and at its history. The case study, in abundant and realistic historical detail, must be the focus of one's inquiries.

But does not this focus trap one in a dangerous circularity? By relying on case studies, is one not testing claims about science by the very methodology which is suspect? Is one not testing and attempting to falsify hypotheses, in a Popperian fashion, even if these hypotheses are about, rather than within, science? I confess to worry that people do not take this problem as seriously as they should—usually because they do not take it at all. My suspicion is that it is not a crippling circularity but rather one of those familiar situations where no ready external point from which to begin inquiries exists. We have to start from where we are. One tests and comes up with results and then, if the results are not to one's liking, one looks at and tests or queries the method.

But whether my suspicion is right or wrong, here at least we can say in turning to science and to its history, we are not playing unfair with the critics of science. This is indeed the very move urged on us by the constructivists. And should things turn out badly for traditional objectivists, it is not a move they can deplore retroactively. Popper always prided himself on the way in which his philosophy reflected the ways and results of actual living science, physics in particular.

Which science to choose? My choice falls on biological evolution-ary theory. I shall look at the whole history of evolutionary theorizing over the 250 years of its existence, from its beginnings in the middle of the eighteenth century down to the work being produced today. Per-haps more than any other area of great science, evolution has been at the focus of the constructivists' attentions. It has been held up as a product of societal thinking rather than a reflection of mind inde-pendent reality. Even some evolutionists think this! "Science, since people must do it, is a socially embedded activity. It progresses by hunch, vision, and intuition. Much of its change through time does not record a closer approach to absolute truth, but the alternative of cultural contexts that influence it so strongly. Facts are not pure and unsullied bits of information; culture also influences what we see and how we see it" (Gould 1981, 21–22). Hence, I am taking a body of work that the critics themselves have taken as paradigmatic for their case. But the proof will be in the pudding.

Because I am not writing history for its own sake, I shall not attempt a comprehensive survey of everything that has happened in the history of evolutionary theorizing. Such surveys do exist (see especially Bowler 1984); I shall concentrate, rather, on a number of key repre-sentative figures in the past and present. What is lost in breadth will, I hope, be gained in depth, and depth is absolutely essential if we are to ferret out the true nature of science.

Values in Science

How is the discussion to be framed? We hardly want the tedium of parading each historical or contemporary figure past all of the various philosophical authorities: Popper, Kuhn, Latour, and others. We need a certain level of generality. The key debate is that between objectivity and subjectivity: Does science obey certain disinterested norms or rules, designed or guaranteed to tell us something about the real world, or is it a reflection of personal preference, the things in culture that people hold dear? In other words, the debate is about interests—the desire for and a devotion to the objective truth, as opposed to an acceptance of (scientists would say a wallowing in) the subjectively social. Interests I take to be another way of speaking of values, and hence I shall take this debate to be one which is crucially about the values of or in science.

The critics—the subjectivists, the constructivists—are clearly arguing about values. Their claim is that science is full of values: sexual values (preferring men over women, or heterosexuals over homosexuals); racial values (preferring Gentiles over Jews, or Jews over Arabs, or whites over blacks); religious values (preferring Protestants over Catholics, or Christians over the heathen); and much more. The subjectivist sees these values as playing a major and noneliminable role in science, sometimes (depending on the circumstances and the particular value) to be deplored, sometimes to be praised and encouraged. Precisely because such values are societal or cultural, they (and the resultant science) cannot tell of a human-free reality: "Theories . . . are not inexorable inductions from facts. The most creative theories are often imaginative visions imposed upon facts; the source of imagination is also strongly cultural" (Gould 1981, 22).

But is the objectivist arguing for values at all? Surely the point is that the Popperian wants nothing to do with values, in science at least. Indeed, in the eyes of some critics, this is part of the problem (Longino 1990, 191):

> The idea of a value-free science presupposes that the object of inquiry is given in and by nature, whereas contextual analysis shows that such objects are constituted in part by social needs and interests that become encoded in the assumptions of research programs. Instead of remaining passive with respect to the data and what the data suggest, we can, therefore, acknowledge our ability to affect the course of knowledge and fashion or favor research programs that are consistent with the values and commitments we express in the rest of our lives. From this perspective the idea of a value-free science is not just empty but pernicious.

In fact, this is a little inaccurate. While it is true that objectivists want nothing to do with the kinds of values just listed above—*cultural* values—they are just as value-committed in their way as the subjectivists are. The ultimate value here is truth, defined as a genuine knowledge of the way that the world actually is. The objectivist values getting things *right*, in the sense of putting one's ideas in correspondence with the reality "out there." Hence, the objectivist also values those ways (methods) that lead to truth. Those norms or modes of reasoning that are thought to put us in touch with reality are generally known as *epistemic values*—a term from the field of epistemology which asks the philosophical questions "How can I know?" "What can I know?" The objectivist claims that science uniquely manifests or obeys those norms.

One might be inclined to use a nontechnical term for the values of the objectivist, a term that can stand in opposition to the phrase "cultural values"—"scientific values" for instance. But there are virtues in using a more neutral term. The debate is about whether the values of science really are truth seeking or are something else again. Hence, at the risk of being accused of being no less jargon-prone than the critics of science, in this book I shall go with the philosophers and adopt the term "epistemic values." I take the major question at issue, therefore, to be: What is the role in science of epistemic values—truth seeking—as against the cultural values, or what we might call nonepistemic values.

Obviously if we are to take our discussion forward, we must unpack in a little more detail the notion of an epistemic value. In particular, we need to list those values which—in the words of the philosopher and historian of science Ernan McMullin—are "presumed to promote the truth-like character of science, its character as the most secure knowledge available to us of the world we seek to understand" (1983, 18). Availing ourselves of McMullin's efforts, we would put high on our list of such values *predictive accuracy:* the power to make forecasts about what one will find in the unknown. Every theory must tolerate some degree of inaccuracy, but overall the theory which does not predict, and do it accurately, is doomed to rejection. The theory which lets us predict suggests to us that it is not just a creation of our imagination but a reflection of something "out there."

We cannot do without the twins of *internal coherence* and *external consistency.* If the parts of a theory do not hang together without contradiction, the theory is discarded. "One recalls the primary motivating factor for many astronomers in abandoning Ptolemy in favor of Copernicus. There were too many features of the Ptolemaic orbits, particularly the incorporation in each of a one-year cycle and the handling of retrograde motions, that seemed to leave coincidence unexplained and thus, though predictively accurate, to appear *ad hoc*" (McMullin 1983, 15). Likewise with the relations between a theory and its fellows. "When steady-state cosmology was proposed as an alternative to the Big Bang hypothesis in the late 1940's, the criticism it first had to face was that it flatly violated the principle of conservation of energy, which long ago attained the status almost of an *a priori* in mechanics" (15).

Unifying power is surely very important for success in science. An excellent example is the geological theory of plate tectonics. "What has impressed geologists sufficiently to persuade most (not all) of them to overcome the scruples that derive, for example, from the lack of a mechanism to account for the plate-movements themselves, is not just its predictive accuracy but the way in which it has brought together previously unrelated domains of geology under a single explanatory roof" (15). And then there is the very significant value of *fertility*. "The theory proves able to make novel predictions that were not part of the set of original explananda. More important, the theory proves to have the imaginative resources, functioning here rather as a metaphor might in literature, to enable anomalies to be overcome and new and powerful extensions to be made" (16).

You will notice that Popperian falsifiability, which was surely intended as an epistemic value or norm, has not made this list; but I take it that it is covered by predictive accuracy, perhaps with an element of coherence or consistency. A theory that is predictively powerful and takes seriously any empirical challenges to its various elements is precisely what we have in mind when we think of a theory as falsifiable. I will not therefore list it separately.

But one other value which I will mention is *simplicity* or elegance—the sense of something aesthetically compelling about a theory. Somewhat uneasily, McMullin lists it among the epistemic values, although admitting that it is "problematic." While not denying its force, deep down inside objectivists fear that simplicity—a Kuhnian favorite—is much more a matter of psychology and taste than of logic. It is a bit too close to the nonepistemic for comfort. For this reason, objectivists have often tried to belittle the significance of simplicity or to translate it into more acceptable epistemic values. Popper (1959), for instance, not entirely successfully, tries to cast simplicity in terms of falsifiability. In the last century, the English philosopher and historian of science William Whewell (1840) rolled together unificatory power and fertility into one value which he labeled a "consilience of inductions," and then argued that the whole package is equivalent to simplicity!

I will not stop here to argue the toss. We are not dealing with an official canon, like the books of the Bible. Rather, we have a set of rules that supposedly are taken seriously as part of good-quality, objective science. We now have the tools for the job—tools which properly reflect the division between (on the one hand) the Popperian objectivist and (on the other hand) the Kuhnian subjectivist.

What Shall We Find?

In opposing epistemic to nonepistemic values, is one not ruling out the fundamental worry that science itself is a product of culture, and thus the epistemic is itself cultural? For now, let me simply say that this is a concern I share and that it will not go unanswered. In the end I hope we can find a way of bringing the epistemic and the cultural together—to marry the strengths of science with the insights of its critics.

Were I writing a detective story, such a hope would suffice as we move right on to the case study. From henceforth, it would be up to you to spot the clues that lead to the denouement at the end of the book. And indeed, if you want to read this way, then do skip right to the beginning of the next chapter. But if you want to follow along, seeing where the argument is going and how such a synthesis emerges, let me sketch out what I believe will be the major findings.

First, without question, the earliest expositions of an evolutionary position were far more the reflection of culture—deeply impregnated with its values—than anything of high epistemic standing. However, then as now, and for much the same reasons, this was considered an unhappy and unsatisfactory state of affairs, and throughout evolution's history the drive has been to replace the cultural with the epistemic. Since this effort has been successful, although it has been a long drawn-out process rather than one of instant rectification, in a sense evolution has been the perfect exemplar of the Popperian philosophy. Objectivism wins!

But not so fast. In another sense, subjectivism wins also. Although epistemic values may push out cultural values, in other respects some cultural values persist—even grow in importance. Not so much as part of the science but rather as values *about* the science; hence I call them *metavalues*. Metavalues both internal to the culture of science (like the desire to publish in the best journals) and external to the culture of science (like the urge to justify certain religious beliefs) serve to reinforce the epistemic values of science. Admittedly, there is no reason why two scientists should share metavalues or why science created by one person could not be accepted by another with different (or no pertinent) cultural values. But we shall find that even where culture is stripped of its values, it can and does persist in science. Not just accidentally either. Culture is the very fabric of science and in itself makes possible the achievement of the desired ends. When two scientists have different cultures, they do have different sciences.

What we shall learn is that a crucial part of the case for subjectivism depends on the significance of scientific metaphor—the transference of ideas from one field to another, most particularly the transference of ideas from some field in the general cultural milieu to a field in the pertinent area of the science. The significance of this transference, incidentally, is very much in line with Kuhn's philosophy. Several years after *The Structure of Scientific Revolutions* he wrote: "Metaphor plays an essential role in establishing links between scientific language and the world. Those links are not, however, given once and for all. Theory change, in particular, is accompanied by change in some of the relevant metaphors and in the corresponding parts of the network of similarities through which terms attach to nature" (Kuhn 1993, 539).

My conclusion therefore will be that indeed both Popper and Kuhn were right about science. In a sense—in the ever-greater exemplification and satisfaction of epistemic norms—evolutionary science is object and aspires toward objectivity. But in another sense—in the unelimimable and significant position of culture, including its values—evolutionary science was and ever remains in the realm of the subjective. And this I take to be a happy conclusion, for when intelligent people fall out, there is usually truth on both sides. However, I recognize that such a balanced view does cause problems for the realism question.

It is one thing to define objectivity and subjectivity in such a way that they can exist harmoniously together, but what about reality? Surely the world is real or it is not? Evolution happened in the way that evolutionists claim—it really did happen in this way—or it did not. Both sides here—Popperian and Kuhnian—cannot be true. One must be false. Yet apparently both sides have merit.

My suspicion is that—as is generally the case when you get two good but conflicting answers to the same question—there is something wrong with the question. Or perhaps with the way in which the question is being answered. Either the whole question of realism versus nonrealism is not a genuine question or, even if the question is genuine, the realism issue—for all that one might think otherwise—cannot be resolved definitively by an appeal to science. And here now I will leave you in suspense, for I am not sure that we can say anything very meaningful on that issue until we have done our work. We shall return to this matter at the end of the book. By then, the picture will appear somewhat clearer. I do assure you that even if we cannot solve the problems that we thought we might solve, we shall have findings pertinent for the solution of equally pressing matters. For now, the time has come to stop the talking and turn to the science and its history.

2

⤬

ERASMUS DARWIN

From Fish to Philosopher

Erasmus Darwin (1731–1802) liked to eat. His hosts knew that he relished a "luncheon-table set out with hothouse fruit, and West India sweetmeats, clotted cream, Stilton cheese, etc." A happy three hours later, the dressing bell sounding, he would express his joy, hoping that dinner "would soon be announced" (Hankin 1858, 1, 154). It is small wonder that at home his table followed the pattern set by the great Thomas Aquinas, with a semicircle cut into its boards so that he might get close to the action.

Darwin would give full return for the hospitality offered. He was one of the most brilliant physicians of the eighteenth century, effecting cures when all others despaired and refusing the entreaties of oft-times crazy King George III only because he did not want to leave his (English) Midlands practice for London and Court life. He enjoyed good conversation and had a capacity for friendship, especially with the great industrialists and scientists of his age, not to mention many men of political thought and action. Benjamin Franklin and Jean-Jacques Rousseau were among his connections (King-Hele 1963).

The ladies, too, appreciated his charms, despite his girth, a stammer, and missing front teeth. A wild youth at the University of Edinburgh medical school convinced him that it would be unwise to follow both Bacchus and Venus, so, becoming a near abstainer, he specialized in love. After his first wife died, leaving him with three children, he set up house with his mistress. Then, two more children later and nearing fifty, he became enamored with the young wife of one of his patients, a retired colonel. This husband conveniently being summoned to the great regiment in the sky, Erasmus Darwin promptly saw off younger rival suitors, married the widow, and fathered another seven children, for a grand total (as far as we know) of twelve.

It is little wonder that Darwin, never slow to put his passions into verse and earning in his day the reputation as one of England's leading poets, should have fallen out of favor with the staid Victorians in the following century. But even if not as a lover or as versifier, his reputation today rides high. We respect him as the first man to express fully, and to argue ardently for, the idea of organic evolution: the idea, as his grandson Charles was to say some 60 or more years later, that all organisms came through a natural-law–bound process of development from "one or a few forms." Let us therefore begin our story by looking at Erasmus Darwin's evolutionism and at the forces behind it. Then, setting the pattern for future discussion, I shall see how his work relates to our concerns.

Everything from Shells

Darwin's evolutionism can be found in several of his various writings, including his poetry; and although never properly presented as a complete scheme in its own right, it is firm and unambiguous. The ideas date from the 1770s or before, when Darwin was much taken by fossil discoveries thrown up by excavations for the Grand Trunk Canal, cutting across the center of the country and connecting by navigable water the east and the west coasts of England. To his good friend the potter Josiah Wedgwood, Darwin wrote excitedly of a trip he had made down into the massive Harecastle Tunnel that was being bored through the Pennines to complete the link: "I have lately travel'd two days journey into the bowels of the earth, with three most able philosophers, and have seen the Goddess of Minerals naked, as she lay in her inmost bowers" (King-Hele 1981, 43).

Not just minerals but fossils also, which Wedgwood was collecting and calling on Darwin's help for identification. Apparently so striking were these finds that Darwin became a transmutationist, even taking as his personal motto *E conchis omnia* (Everything from shells), which he shortly had painted on the door of his personal carriage. Some twenty years passed, however, before Erasmus Darwin started to put these thoughts down on paper, in systematic form, breaking explicitly into print in his major medical treatise, *Zoonomia*.

Not really that systematic, actually, for Darwin made little attempt to disentangle the various aspects of his thinking. Today's evolutionists usually distinguish three parts to their theorizing. First are claims about the very *fact* of evolution—that all organisms descend through natural causes (that is, no miracles) from life forms very different from themselves, perhaps ultimately from one or a few very primitive life forms. Second are claims, or questions, about what *paths* (or phylogenies) organisms took on their journey to the present. Did the birds come from dinosaurs, for instance, or did they evolve straight from earlier reptiles? Third are claims, or questions, about the *causes* or mechanisms of change. Were there many causes or just one prime one? Erasmus Darwin jumbled up his thoughts about these three issues, and what starts out as an argument for the fact of evolution will end with a conclusion about the causes of evolution, and so on (Darwin 1794, 500–505).

The most direct arguments for the fact of evolution were, first, the analogy that Darwin saw between evolution and individual development ("from the feminine boy to the bearded man, and from the infant girl to the lactescent woman") and, second, the similarities between the bodies of organisms of different species (see figure). The latter, known today as homologies, were taken by Darwin to be evidence of common ancestry. To all of this, presumably, Darwin added the existence of fossils, for certainly they did not figure prominently in any discussion of pathways. No attempt was made to trace life's history in the rocks. Indeed, about all that Darwin said on this matter at this point was that whales, seals, and frogs—those animals which bridge land and sea—seem to hold clues as to the route taken by life in its development. Of most interest to him was the question of causes, and all sorts of bits and pieces of anecdote and hearsay were offered up here.

Coming as he did from the agricultural part of England, Darwin was naturally much taken with the breeding of horses, dogs, and cattle.

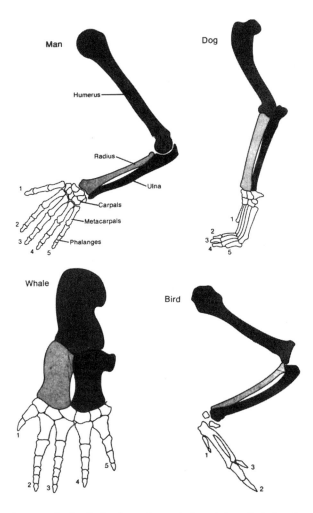

Homologies between the forelimbs of vertebrates (adapted from Dobzhansky et al. 1977).

He was sure that there was significance in the "great changes introduced into various animals by artificial and accidental cultivation." Also, Darwin wrote of the changes that are brought on first artificially by the natural environment or by human action but later get ingrained in heredity. He pointed to the docking of dogs' tails, which (he believed) leads eventually to animals born naturally with little or no tail. Paradoxically, this very old belief in the inheritance of acquired characteristics—characteristics that the organism presumably acquired through use or disuse of certain body parts or behaviors—is now known as Lamarckism after the French evolutionist Jean-Baptiste Lamarck, even though his writing came at least a decade after Erasmus Darwin's.

Finally, Darwin was fascinated by the idea that changes can come from within by virtue of personal desires and needs. This is connected to a psychological movement known as "association," promoted by the eighteenth-century thinker David Hartley, who argued that habits and experiences can lead to new beliefs. Extending the psychological to the physiological, Darwin argued explicitly that "I would apply this ingenious idea to the generation or the production of the embryo, or new animal, which partakes so much of the form and propensities of the parent" (480). This notion was backed by a mechanical hypothesis that small particles are produced by bodily parts, carried by the blood to the sex organs, and whence combined by both parents to make new offspring. (A similar view would be held by grandson Charles.)

One cannot honestly say that Erasmus Darwin was overly concerned about the minutiae of the evolutionary process. Take adaptation, for instance, a question of some interest if only because of the extent to which it was to obsess Charles. No one—especially no one in England writing at the end of the eighteenth century—could have been unaware of the significance of adaptation. For all that David Hume in his *Dialogues Concerning Natural Religion* (1779) had offered a devastating critique of the so-called argument from design—which attempts to prove the existence of God by arguing that design in the natural world implies a Grand Designer—that particular brand of theological reasoning throve as never before. Erasmus Darwin wrote only a year or two in advance of Archdeacon Paley's definitive tome, *Natural Theology*, the work which likens the eye to a telescope, arguing that as the telescope requires a telescope maker so the eye requires an eye maker or Maker. *The* mark of the living was its adaptiveness, its end-directed or teleological nature, the fit between structure and function.

Although certainly Erasmus Darwin acknowledged adaptation—"The trunk of the elephant is an elongation of the nose for the purpose of pulling down the branches of trees for his food, and for taking up water without bending his knees" (504)—it was not the driving force behind his vision of the organic world. For Erasmus Darwin, what counted was the big picture, a view which in his later writings he was happy to express in florid verse (Darwin 1803, 1, 295–314):

> Organic Life beneath the shoreless waves
> Was born and nurs'd in Ocean's pearly caves;
> First forms minute, unseen by spheric glass,
> Move on the mud, or pierce the watery mass;
> These, as successive generations bloom,
> New powers acquire, and larger limbs assume;
> Whence countless groups of vegetation spring,
> And breathing realms of fin, and feet, and wing.
>   Thus the tall Oak, the giant of the wood,
> Which bears Britannia's thunders on the flood;
> The Whale, unmeasured monster of the main,
> The lordly Lion, monarch of the plain,
> The Eagle soaring in the realms of air,
> Whose eye undazzled drinks the solar glare,
> Imperious man, who rules the bestial crowd,
> Of language, reason, and reflection proud,
> With brow erect who scorns this earthy sod,
> And styles himself the image of his God;
> Arose from rudiments of form and sense,
> An embryon point, or microscopic ens!

Values in Erasmus Darwin's Science

We turn now to analysis. First, recognizing that thus far these values have been more stipulated than justified, how far would one want to say that Erasmus Darwin's ideas were governed or shaped by epistemic constraints such as predictive ability, consistency, and coherence? One hardly need pause before responding: not in any great degree.

It is true that Erasmus Darwin was trying to make sense of the facts of nature as he saw them. He was not simply making everything up out of his mind. He was offering something which is a candidate for science, unlike, say, the products of the imagination of a novelist. Moreover, it would be ungenerous to claim that he was indifferent to the virtues of a unified, integrated picture. Indeed, Darwin should be commended for bringing so much—observation, anecdote, hearsay, conventional wisdom—beneath the umbrella of organic change. Most importantly, Darwin was proposing a picture based on natural law rather than miracle—a precondition for the realization of epistemic factors.

But after granting all that, we have to admit that the call of the epistemic itself is faint. There is little possibility of, or attempt at, prediction. Apart from some general debts to psychology, coherence and consistency are absent. Practically nothing can be found to suggest that his ideas have the fertile power to push into new areas of inquiry, producing in turn science of a high epistemic standard. And simplicity is not a notable virtue of Erasmus Darwin's thinking.

I do not want to imply that this is a set of failures that should be laid exclusively at the feet of Dr. Darwin. Geology in the eighteenth century, for instance, was hardly in the developed state that it had entered by the time his grandson published (Laudan 1987). The fossil record was unknown, certainly in any integrated fashion; Erasmus Darwin could not be expected to be consistent or inconsistent with a subject that was in a virtually nonexistent state. However, the fact remains that whatever the assignment of praise or blame, the evolutionism of Erasmus Darwin did not rate high in the epistemic stakes.

If not epistemic, one is led at once to ask about the nonepistemic, cultural side of his thinking. Here the story is very different. Start with religion. Erasmus Darwin was no Christian. Like many in his age, he believed in an Unmoved Mover—a God who had set things in motion and then stepped back from his handiwork. He was a deist, a person who—as opposed to the theist (traditionally Christian, Jew, or Muslim)—sees the highest mark of God's power and glory not in divine interventions (that is, miracles) but precisely in the fact that miracles are not needed. God can do everything through unbroken law. To use a modern metaphor, he has preprogrammed the world so that further intervention is unnecessary. Evolution, the triumph of unbroken law, can therefore be seen as the apotheosis of God's standing and worth. Everything is planned beforehand and goes into effect through the laws of nature. In Darwin's own words: "What a magnificent idea of the infinite power of *The Great Archi-tect! The Cause of Causes! Parent of Parents! Ens Entium!*" (1794, 509).

I am not sure that any of this, in itself, makes Darwin's theory a religious theory, whatever that might mean. Atheists and agnostics, or for that matter certain types of Protestant Christians, are no less eager to see the world working according to unbroken law (Barbour 1988). But inasmuch as Darwin's belief in the law-bound nature of the world relied heavily on deism for support and plausibility (which it certainly did), one might fairly say that a cultural (nonepistemic) value of religion was a factor in Darwin's theorizing. In the next chapter, when dealing with the similar religious ideas of Charles Darwin, I will explore the precise way in which religion *qua* culture operated in the grandson's scientific work. Here, I want to push a little further our recognition that religion was but one part of the cultural indebtedness we find in Erasmus Darwin's thinking on evolution.

I have just written that programming is a modern metaphor. In fact, this is not quite accurate. Even at the time of Erasmus Darwin, people were starting to think in such terms, specifically in the weaving industry, where mechanical methods (complete with punch cards) were devised to set looms to weave predetermined patterns. This reached a climax in the 1830s when, using this technology, Charles Babbage attempted (ulti-mately unsuccessfully) to build an elementary computing machine. Significant was the fact that Babbage's machine could produce series of numbers with anomalies built in: 1 to a million, and then 1,000,002 rather than 1,000,001, for instance. Immediately, Babbage (1838) saw this work as having theological implications. Apparently God, the supreme ma-chine maker, could have preprogrammed the world to produce or do anything He wanted, through unbroken law. Even though we humans might regard such a strange happening as miraculous, it would not in fact have required special intervention from above.

This was a somewhat extreme exemplar of the idea of God as machine maker, but in England, the home of the Industrial Revolution, it was an exemplar and not an anomaly. Thus the Reverend Baden Powell (1855, 272), Savillian Professor of Geometry at Oxford (and, incidentally, father of Robert Baden-Powell, founder of the Boy Scouts), had this to say:

Precisely in proportion as a fabric manufactured by machinery affords a higher proof of intellect than one produced by hand; so a world evolved by a long train of orderly disposed physical causes is a higher proof of Supreme intelligence than one in whose structure we can trace no indications of such progressive action. And in proportion as we might be able to follow out more and more details of that succession of causes, should we derive increasing evidence of the great truth.

This industrial metaphor was central to Erasmus Darwin's thinking about evolution. His deism was part of a package deal which saw universal excellence in the local world he inhabited—excellence which came about because of the changes that Darwin and his scientist and industrialist friends (often the same people) were effecting. Britain was being changed: energy, coal and steam primarily, was being harnessed; efficient methods of transportation, especially canals, were being built; the population was moving to cities to work in factories for the efficient use of labor; new machines were being invented and produced; minerals were being excavated and utilized; the rudiments of political economy were taking shape; and much more. Darwin resided at the geographical heart of all this activity, the British Midlands, and he was linked to the social nerve network as a member of a monthly dining club, the Lunar Society, which included Josiah Wedgwood, the chemist Joseph Priestly, and the industrialist Matthew Bolton. Erasmus Darwin celebrated Albion's achievements in prose and verse, above all the central belief that through human effort, ingenuity, and daring we can make an ever-better society (Schofield 1963). Forward means better, better means happier, and happier means progress, limitless progress.

If progress requires a break from the old power structure, so be it. Theologically, progress has generally been taken as antithetical to Christianity, inasmuch as Christianity stresses the providential intervention of the Creator. Without His miraculous aid, we, unaided, can do nothing. For Erasmus Darwin this break with Providence was more than just a matter of theology; it was the beginning of a whole new philosophy of life. The faster England could be torn away from its slumbering rural roots, its gentry landowning establishment, its Church-dominated oppression, and transplanted into a middle-class, science-loving, democratic-thinking, rationalist-inclined, urban-based meritocracy, the better.

And part and parcel of this philosophy is a belief in evolution—an evolution whose path is upward-looking, organic-improving, human-directed, progressive. Indeed, evolution "is analogous to the improving excellence observable in every part of the creation; such as in the progressive increase of the wisdom and happiness of its inhabitants" (Darwin 1794, 509). Truly might one say that the evolutionism Erasmus Darwin espoused was the industrialist's philosophy made flesh: From "An embryon point, or microscopic ens!" to "Imperious man, who rules the bestial crowd."

In good, strong, circular fashion, Darwin started with his social belief in the desirability of progress—the progress of the British industrialist. He read this into nature, and then he read it right back out as confirmation of his philosophy. "All nature exists in a state of perpetual improvement . . . the world may still be said to be in its infancy, and continue to improve FOR EVER and EVER" (Darwin 1801, 2, 318).

The conclusion has to be that Erasmus Darwin's evolutionism was thoroughly culturally laden. Deism bound with and leading to progress was what motivated Darwin and was the force behind his speculations on organic origins. The epistemic played a very secondary role, if that. Or, putting matters another way, one cannot find very much objective about the kind of work Erasmus Darwin produced. It was as subjective as a Sunday sermon to the faithful or an election address to the voters.

Does this mean, then, that the social constructivist wins, almost before we begin? Hardly, at least in the eyes of the objectivist. Surely, such a person will say, we must inquire into the status of Erasmus Darwin's evolutionary theorizing. *We,* after all, are talking about real science, good science. Not necessarily the greatest science which has ever been produced, but certainly science of top quality. Mature science or science on the way to maturity. Can one truly characterize Darwin's work in these terms? If not, then at worst it tells us nothing and at best it invokes some compromise position: science starts in subjectivity but over time, with continued observation and experiment, with continued application of the epistemic rules, science moves into objectivity. Evolution may have started in culture, but this does not mean that through time we shall see no emergence from culture into epistemic purity.

So how do we rate Erasmus Darwin's work? Do we think of it as real science—good science—or not? Unfortunately, if we stay just within our own terms, the argument will at once run into opposition based on *a priori* suppositions. The objectivist will deny the status; the subjectivist will affirm it. The one will say it cannot be good science, because of its cultural values. The other will say that cultural values are precisely what one expects to find in all science, good or bad. To cut this knot, we must now turn to the past. If everybody at the end of the eighteenth century was perfectly happy with Darwin's work, giving it full scientific status, then the objectivist must at least explain how standards of epistemic necessity have changed. Conversely, if the work was given less than five-star billing—judged as science, by Darwin's own contemporaries—then the objectivist is alive to fight another day. No proof has been offered to show that good science never has cultural values, but the case is not closed almost before discussion is begun.

Erasmus Darwin's Evolutionism in His Own Time

How did his contemporaries rate Erasmus Darwin's work? Was the work thought of as science, as good science, that is? Or was it rated less highly, as mere proto-science or even worse, as pseudo-science? We can answer these questions with some confidence, because the end of the eighteenth century and the beginning of the nineteenth was the time when the notion of a professional scientist was really beginning to solidify—when one knew what was expected of someone who wanted to produce good-quality science and one had standards by which to evaluate such science.

At the highest organizational level were national bodies such as the French Academy of Sciences; the British Royal Society at that time did not have such high status. More locally and specifically, limited discussion groups or work groups formed around certain key figures, for instance the Paris-based chemical Société d'Arcueil, which was associated with the French scientist Berthollet. At the individual level, one had men like Priestley and Black in England and Lavoisier and Laplace in France; then, at the beginning of the new century, Thomas Young and John Dalton in England, Jean Fresnel and Georges Cuvier in France (Ruse 1996).

When we look carefully at what people actually did, empirically and theoretically, in science, we find ourselves in fairly familiar territory, epistemically speaking. Everyone agreed that one must be careful about methodology, in measurement and so forth; that predictive accuracy counted highly (an increasingly important source of funds were governments who were finding that they could use predictive science in such areas as warfare); that being consistent in what one did both within one's own science and with other sciences was important; and that if a science pointed the way to new areas of study with exciting questions and stunning answers, this was a high mark of favor.

Contentious was the question of hypotheses, especially supposedly high-powered, sweeping hypotheses covering and answering many questions at once. The great Newton had said "Hypotheses non fingo," and while there was debate as to the exact meaning of this somewhat cryptic statement, everyone agreed that it meant in some sense hypotheses were a bad thing. One ought rather to be "inductive," where presumably this meant staying close to the facts and gathering many of them, letting the truth emerge naturally (Laudan 1981). One should try to remain in touch with the immediately sensed, in some fashion.

But this presumption of empiricism was not an absolute dictate. In real life—in real science—hypotheses were acceptable so long as they led to new predictions, especially if at the same time they unified lots of disparate areas of experience. The newly conquering wave theory of light was rightly seen as magnificent science, and anything less inductive in the traditional sense it would be hard to imagine. Wave theory was accepted because it did so very well on all of the other cherished epistemic values. People knew good science when they saw it (Buchwald 1989).

Hence—a point to bring joy to our objectivist—the epistemic criteria by which people in Erasmus Darwin's day judged science "good" were essentially identical to the criteria by which people (objectivists, at least) judge science "good" today. Then as now, the things that counted were predictive accuracy, coherence, consistency, unificatory power, fertility, simplicity. Moreover, professional scientists could spot pseudo-scientific systems whose claims are culture-driven in gross violation of epistemic norms; today we might put on that list scientology, astrology, and Nazi racial theory. This negative judgment could be made even when—especially when—one was dealing with something wildly popular with the general public.

A case in point in Erasmus Darwin's day was Mesmerism, a medical hypothesis which involved sweeping claims about animal magnetism, its disorders, and the possibilities of near miraculous cures for sickness (Darnton 1968). In exasperation at the divide between professional scientific skepticism and public enthusiasm, the French king Louis XVI had appointed a commission (chaired by Ben Franklin and including Lavoisier) to look into Mesmer's claims. Back came a withering analysis, showing that none of the purported effects could be justified. Its success rate was no more than that of a placebo. Mesmerism led to no new findings, to no new ideas, to no useful connections with conventional medicine. In fact, it led to nothing at all, save hysteria and danger (Franklin et al. 1996, 83):

> The Commissioners, having recognized that this Animal-magnetism fluid cannot be perceived by any of our senses, that it had no action whatsoever, neither on themselves, nor on patients submitted to it; having certified that pressure and touching occasion changes rarely favorable to animal economy and perturbations always distressing in the imagination; having finally demonstrated by decisive experiments that the imagination without magnetism produces convulsions, and that magnetism without imagination produces nothing; they have unanimously concluded, on the question of the existence and utility of magnetism, that nothing proves the existence of Animal-magnetism fluid; that this fluid with no existence is therefore without utility; that the violent effects observed at the group treatment belong to touching, to the imagination set in action and to this involuntary imitation that brings us in spite of ourselves to repeat that which strikes our senses.

With the commissioners' final epistemic condemnation ringing in our ears—"all group treatment in which the means of magnetism will be used, can in the long run have only disastrous effects"—we have an appropriate point to return to Erasmus Darwin's evolutionism, for it is clear that people viewed his work as being very much at the Mesmerism end of the scale. Like us, they could see that he was in no sense quantitative or experimental. He made no systematic study of nature, for instance the fossil record. Although he certainly drew on the psychology and medicine of the day, he made no effort at genuine integration with the known science. There was no prediction, or anything else very much except a string of anecdotes and stories and reports of curious facts—something more suited for Ripley's *Believe It or Not* than for serious science.

One key passage, for example, told of a man with one dark child among a family of fair children. While his wife was pregnant, the father apparently became obsessed sexually with the dark-haired daughter of one of his tenants. He offered her money, but she spurned his advances. Yet, he had to admit that "the form of this girl dwelt much in his mind for some weeks, and that the next child, which was the dark-eyed young lady above mentioned, was exceedingly like, in both features and colour, to the young woman who refused his addresses" (Darwin 1794, 523–524). About at the same level was the information from the past: "the phalli, which were hung round the necks of the Roman ladies, or worn in their hair, might have effect in producing a greater proportion of male children" (524). Small wonder that even Darwin's supporters worried that his argumentation was more charming than definitive. "If Doctor Darwin had indulged less in theory, and had enlarged the number of his facts, our satisfaction would have been complete" (anonymous writer in the *Monthly Review*, 1800, quoted in McNeil 1987, 174).

Working biologists tended to be scathing on the subject of speculations like Darwin's. They were already worried about the status of their science: for too long it had been merely charming and discursive natural history, no match for the physical sciences. Now the evolutionists were showing that the practitioners of the more established sciences may have some good grounds for their contempt. Tensely and obsessively the orthodox recited the mantras of good science: "Experience alone, precise experience, made with weights, measures, calculations and comparisons with all the substances used and all the substances one can get, that today is the only legitimate way to reason, to demonstrate" (Cuvier 1810, 390).

Of course, the real clue to the low status of Darwin's evolutionism lies in the fact that so much of it appeared in verse. This was no way to do serious science. Indeed, Darwin himself recognized this, to the extent that at first he published anonymously lest he ruin his reputation as a physician. He was presenting an overall picture to the general public—the philosophy of an industrialist, a philosophy of progress, of which evolutionism was a part, not so much a scientific theory in its own right but an element in a world picture. And it is significant that this is precisely how it was treated.

Erasmus Darwin's real critics were not serious scientists but rather those who opposed his philosophy. At the end of the eighteenth century, the excesses of the French Revolution—which Darwin had supported initially—sickened and frightened the English, especially those in power. Could what happened in France somehow cross the Channel, as cattle plague often did (Ritvo 1987)? Shoring up Britain's defenses, the traditionalists took on their opponents. Darwin's reputation was savaged by a wicked parody of his poetry by three of the leading conservatives, including George Canning (Canning, Frere, and Ellis 1798). By the time they had finished, Darwin was a laughingstock and his philosophy was in tatters.

People were very wary indeed of progress—and that went for evolution, too. It was not so much that everybody wanted to go rushing back to a biblically based story of origins; they just wanted to stay as far away as possible from such life- and property-threatening doctrines as the deism-based, progressionist philosophy of Dr. Erasmus Darwin.

Into the Nineteenth Century

We have made a start, although hardly enough yet to draw well-grounded conclusions. If by pseudo-science one means work that is driven by cultural values to the detriment and virtual exclusion of genuine attention to epistemic constraints, then evolutionism at the time of its birth toward the end of the eighteenth century was a pseudo-science. There is no disproof of the objectivist here, nor is there support for the subjectivist. Matters are still unresolved. We must therefore start to move the clock forward—something our history urges us to do, for although Erasmus

Darwin may have been the first major evolutionist, he was not the only one. Very important in the history of the idea is the Frenchman Lamarck, whose *Philosophie Zoologique* was published in 1809 (Burkhardt 1977). And there were others, especially a host of evolutionists or near-evolutionists in Germany. The idea did hang fire through the early decades of the nineteenth century, owing not just to religious and philosophical objections but to the scientific criticisms of Cuvier (1813), who pointed to gaps in the fossil record, the failure of organisms to change since the time of the pharaohs (as inferred from the mummies), and the impossibility of changing one domestic species into another. But by the 1830s we find more and more people starting to speculate in new directions on the question of organic origins—what the British astronomer and philosopher John F. W. Herschel labeled the "mystery of mysteries" (Cannon 1961).

Even the religious kept rubbing away at the matter. People took the Bible seriously. Theists would have said (sincerely) that they took it as true. But what that meant exactly was another matter. The evidence from geology was ever present and growing, particularly in England, where this particular science had such great economic importance: the search for fuel, for minerals, for the best way to build efficient paths of transportation (changing in the nineteenth century from canals to railways). Six literal days of creation, six thousand years ago, were simply not credible. And so more science-friendly interpretations of the origin of species became the norm.

Not that we should at once assume that people were truly sliding straight toward a full-blown evolutionism, or for that matter toward any kind of genuine naturalistic (that is, regular-law–bound) solution. The Bible (revealed religion) might have been open to new readings and in that respect flexible, but natural theology (getting at God through reason and the senses) continued to have negative things to say on the subject of origins. The 1830s saw the very apex of the Argument from Design, with the publication of the eight *Bridgewater Treatises,* volumes whose express intent was to show God's designing purpose in His creation (Gillispie 1950). And the way to do this was by stressing organic adaptation—the hand and the eye—and no one could see how blind, unguided law could lead to such intricacy. In the end, therefore, reason seemed on balance to disprove evolution, rather than support it.

Yet the question of origins was a worry. No dispute about that, for all the confident talk otherwise. William Whewell concluded that science says nothing but "she points upwards" (1837, 3, 588). Which is all very well, but not much of an answer, at least, not much of a very satisfactory answer to a working scientist. An ambitious newcomer might well see an attractive research program here—as we shall now learn.

3

>≤

CHARLES DARWIN

On the Origin of Species

Dr. Robert Darwin was not pleased. His younger son, Charles, had left Edinburgh University and medical training for Cambridge University and the prospects of the Anglican ministry. Now, in August 1831, the lad was proposing a several-year delay before he took orders, so that he could board the navy ship *HMS Beagle* for a trip to South America, as a kind of unpaid companion to the captain. Reluctantly, Robert agreed that the trip could go ahead, but only if Charles could win the independent approval of someone whom the Darwin family respected. Charles Darwin did win such approval—that of his Uncle Josh, son of Josiah Wedgwood, the potter and friend of Erasmus Darwin. And so the young Darwin spent the years from 1831 to 1836 on a voyage which ended by circumnavigating the globe, an experience that led directly to his becoming an evolutionist.

To be fair to Robert Darwin, there were good reasons for dismay. Already, his older son, Erasmus, was establishing a life-long pattern of nonactivity, preferring the literary and social life of London to the demands of steady employment. Now the younger son, Charles, seemed to be going the same way, dabbling in natural history and insect collecting. But Uncle Josh, who thought that the *Beagle* voyage would be character-building, was right. Even at Cambridge, Charles had started mixing with the leading scientific men of the day, and the time away from England hardened his resolve to make his own major contribution to our understanding of nature. This determination persisted throughout his life, producing in 1859 the classic *On the Origin of Species by Means of Natural Selection.*

Darwin's Science

Charles Darwin began his scientific career as a geologist, fired by the newly published *Principles of Geology*, in which Charles Lyell argued that all of the processes of mountain building and continent formation and the like are the end results of slow-working but long-lasting causes of a kind still operative today. This led Darwin to an interest in biogeography, the study of the geographical distribution of plants and animals—a key to the understanding of past geological formations. Thus directed, he was naturally puzzled by the strange distributions of the organisms on the Galapagos Archipelago, which his ship visited in 1835. Why should the birds and reptiles be similar but different from island to island, especially when members of one species might range the whole length of South America?

Soon after young Darwin returned to England, the bird taxonomist John Gould persuaded him that these differences really did add up to different, though similar, species. And so, seeing no other reasonable option, Darwin slipped over into evolutionism. He concluded that the animals had originally come to the Galapagos from the mainland and, once there, had diversified from island to island, owing to the geographical isolation of the islands from one another. For Darwin, this kind of "transmutation" became the key to the whole variety of organic life, past and present.

A number of factors were on Darwin's side here, urging him on into evolutionism. Certainly family influences counted. Grandfather Erasmus had died before his grandson was born, but the tradition was known, and Charles had read the key works, especially *Zoonomia*. More than that, although his undergraduate intention of a clerical career was based on a literal reading of the Bible, by the end of the *Beagle* voyage Darwin's faith had matured into some form of deism, the traditional belief of his family. Although still firmly committed to the existence of God, Darwin looked for His power in unbroken law rather than in miraculous intervention. As was true for his grandfather, evolution was a confirmation of Charles Darwin's religious position rather than an anomalous belief in need of explanation.

Yet evolution had so far failed to explain design or adaptation at the individual level—and this failure was the major reason why most people found evolution simply impossible to accept. That problem had still to be solved. Not that Darwin ever doubted or ducked this obligation, for as a graduate of early nineteenth-century Cambridge University he was not only sensitive (in a way quite missing in his grandfather's thinking) to the functional nature of organic characteristics, but he knew full well what had been the contribution of his alma mater's most famous scientific son. Isaac Newton's greatest achievement had been the concept of gravitational attraction, the ultimate *causal* explanation of the motions of the planets observed by Copernicus, Kepler, and others during the Scientific Revolution. If Darwin was to become the Newton of biology—and this was certainly his hope—then he too had to provide causes. To argue simply for the *fact* of evolution was not enough. One had to say what it was that made evolution work.

Here, as in so much, Charles Darwin was in a favored position. Reaping the benefits of the Industrial Revolution, his family was able to play at being agriculturalists, and through his connections Darwin soon learned that the secret to animal and plant breeding does not lie in Lamarckian types of inheritance but in selection: taking the best and using that as the breeding stock (Barrett et al. 1987). Picking or choosing in some sense had to be the key to change. The problem was to see how this could apply in nature.

The crucial insight came through a reading of a well-known tract on political economy, *An Essay on the Principle of Population*, by the Reverend Thomas Robert Malthus. Turning on its head that work's gloomy claims about the impossibility of permanent change because of the pressures of ongoing population numbers on limited space and food, Darwin saw that in the consequent "struggle for existence" he would have a motive force behind a nonhuman form of breeding. He would have a mechanism for evolutionary change, so-called natural selection—a mechanism, moreover, which speaks to the adapted nature of organisms.

This is the way natural selection works: Many more organisms are born than can possibly survive and reproduce; organisms come with heritable differences; those organisms that succeed in the struggle for existence and reproduction will be different from those that do not, and their success will (on average) be a function of the differences; the organisms that succeed will pass their characteristics along to offspring, whereas the organisms that do not succeed will not; hence, there will be an ongoing process of winnowing or selection of organisms with adaptive characteristics; and given enough time this will lead to full-blown evolution. The hand and the eye are indeed adaptive, but they came to be that way through no special Divine intervention. In His design of the living world, as in His design of the planets, God works through the processes of unbroken natural law.

A mechanism is not a theory, however. Darwin discovered—let us not prejudge issues; Darwin formulated—the concept of natural selection in the autumn of 1838, and for the next few years he worked hard to put his ideas into a fully developed theory. But although he wrote these down in the early 1840s, for reasons still not completely understood he became diverted by a massive study of barnacle taxonomy (Darwin and Wallace 1958). It was only at the end of the 1850s, after a young naturalist, Alfred Russel Wallace, had also hit upon the idea of natural selection as a force for evolutionary change that Darwin finally published his ideas. At once recognized for the important work that it was, the *Origin of Species* was to go through six editions in the next twelve years, as Darwin responded and revised in the light of criticisms and new findings.

One Long Argument

"When on board H.M.S. 'Beagle,' as naturalist, I was much struck with certain facts in the distribution of the inhabitants of South America, and in the geological relations of the present to the past inhabitants of that continent. These facts seemed to me to throw some light on the origin of species—that mystery of mysteries, as it has been called by one of our greatest philosophers" (Darwin, 1859). Let us look for a moment at the *Origin* itself and then go on to put it in context.

Darwin opened by talking about artificial selection, both to prepare the reader for natural selection and to start building his evidential case. Then on to natural selection, something which in later editions he was also to term the "survival of the fittest" (language originally used by the English philosopher, sociologist, and biologist Herbert Spencer). Also, Darwin introduced a secondary mechanism, sexual selection, which involves a struggle less for food and survival and more for mates and reproductive success. Darwin also accepted other secondary mechanisms which no evolutionist would accept today, including the Lamarckian inheritance of acquired characters. (In the language of evolutionists, characters means, loosely, characteristics, which may be either physical or behavioral or, commonly, both.)

Closely related to the discussion on selection was Darwin's causal thinking about speciation. Believing that the real changes in evolution come when a lineage splits into two (as obviously happened in the Galapagos), Darwin spoke of his "principle of divergence," which apparently comes about when organisms divide up the spoils; selection promotes the splitting and multiplication of types so that resources can be more efficiently exploited. This led Darwin to his famous description of life's history as akin to a magnificent tree (1859, 129–130; see figure):

> The affinities of all the beings of the same class have sometimes been represented by a great tree. I believe this simile largely speaks the truth. The green and budding twigs may represent existing species; and those produced during each former year may represent the long succession of extinct species. At each period of growth, all the growing twigs have tried to branch out on all sides, and to overtop and kill the surrounding twigs and branches, in the same manner as species and groups of species have tried to overmaster other species in the great battle for life. The limbs divided into great branches, were themselves once, when the tree was small, budding twigs; and this connexion of the former and present buds by ramifying branches may well represent the classification of all extinct and living species in groups subordinate to groups . . . As buds give rise by growth to fresh buds, and these, if vigorous, branch out and overtop on all sides many a feebler branch, so by generation I believe it has been with the great Tree

PEDIGREE OF MAN

The evolutionary tree of life as envisioned by one of Darwin's German followers, Ernst Haeckel (1896).

of Life, which fills with its dead and broken branches the crust of
the earth, and covers the surface with its ever branching and
beautiful ramifications.

Of course if selection is to be effective in bringing about the evolution of new species, one needs both a source of new variation and a
mechanism for that variation to be passed on to offspring. To be candid,
Darwin had little of positive value to say on this score, and we get
something of a jumble of suggestions for how variation comes about,
many of a Lamarckian nature. Then, following a discussion of difficulties
in his theory (intermediate forms, highly adapted features, and the like),
it was with some relief that Darwin turned to the second major part of the
Origin. This was a survey of the various branches of biology, showing how
evolution through selection throws light upon them and conversely how
collectively they support and make probable evolution through selection.

Instinct and behavior were major topics of interest, especially as
they apply to the social insects like ants and bees. Darwin put much care
into showing how the instinct and behavior of social insects would fall
under selection no less readily than would physical characteristics. "No
one will dispute that instincts are of the highest importance to each
animal. Therefore I can see no difficulty, under changing conditions of
life, in natural selection accumulating slight modifications of instinct to
any extent, in any useful direction" (243).

Not that the physical was ignored by Darwin. Geology and paleontology in particular got detailed discussion. He devoted some time to
the incompleteness of the fossil record, noting that one simply would
not expect to see that many intermediate forms. Much more time was
spent on the positive facts in favor of evolution, for instance that the
more general and linking forms are found lower (and hence earlier) in
the fossil record, whereas the more specialized forms come higher (and
hence later). Also, time was spent on facts that are puzzling if one
subscribes to some form of special creation but not at all if one is a
(Darwinian) evolutionist. Why, for instance, does the fossil record tell
us that when once an organism has gone extinct, it never reappears at a
later date? Simply because life never moves back on itself in any
significant way.

Geographical distribution was one of Darwin's strongest suits, as might be expected given his own experiences in the Pacific, particularly in the Galapagos archipelago. Then before a summary chapter, Darwin moved into a kind of grab-bag discussion of classification, morphology, embryology, and rudimentary organs, showing how each of these topics supports the overall case and in turn is given its own place and explanation in the biological scheme of things. Embryology particularly had always been Darwin's pride, as he argued that the reason for the frequent similarity of the embryos of organisms very different as adults lies in the fact that natural selection acts only on the adults. The selective forces operating on embryos are always the same and hence there is no reason for the forms to evolve apart.

Here particularly Darwin showed how significant for him was the analogy between the domestic and natural worlds; this played a crucial role both in his exposition and in the evidentiary support he brought to the argument. Breeders select for the adult, and so the characteristics of juveniles are generally much closer together (despite what people might think): "Fanciers select their horses, dogs, and pigeons, for breeding, when they are nearly grown up: they are indifferent whether the desired qualities and structures have been acquired earlier or later in life, if the full-grown animal possesses them" (446). Likewise, natural selection in the great world outside selects adult characters.

And so the case was brought to completion. Darwin accurately described what he had done as "one long argument" from beginning to end: "From the war of nature, from famine and death, the most exalted object which we are capable of conceiving, namely, the production of the higher animals, directly follows. There is a grandeur in this view of life, with its several powers, having been originally breathed into a few forms or into one; and that, whilst this planet has gone cycling on according to the fixed law of gravity, from so simple a beginning endless forms most beautiful and most wonderful have been, and are being, evolved" (490).

Epistemic Values in Darwin's Science

To begin our analysis of Darwin's science, let's start with the value of predictive accuracy—the extent to which the theory of the *Origin* gave evidence of ability to project into the unknown. At one level, the theory comes across as rather disappointing. Certainly we find no long-term predictions, such as what will happen in the next million years to the elephant's trunk. Nor is there much in the way of short-term predictions, at least not of a quantitative nature. Darwin does not give any experimental reports of, say, how different populations perform under different (artificially controlled) selective pressures. This sort of thing is simply not there.

However, it would be wrong to conclude that there is nothing of a predictive nature in the *Origin*, especially if one understands (as one surely should) prediction in the more general sense as covering not merely prediction of future events or phenomena but prediction of current events or phenomena that are nevertheless unknown to the writer at the time that the theory is formulated. Darwin's first example of selection in action—a hypothetical example—is the predator–prey interaction of wolves and deer, with different wolves having different strategies: some fast, some strong, some crafty. Suppose now that for some reason the proportion of fast deer increases: "I can under such circumstances see no reason to doubt that the swiftest and slimmest wolves would have the best chance of surviving, and so be preserved or selected,—provided always that they retained strength to master their prey at this or at some other period of the year, when they might be compelled to prey on other animals" (90).

This is certainly a prediction in a hypothetical situation, and indeed Darwin goes on then to suggest that in some situations it, or something much like it, actually obtains. "According to Mr Pierce, there are two varieties of the wolf inhabiting the Catskill Mountains in the United States, one with a light greyhound-like form, which pursues deer, and the other more bulky, with shorter legs, which more frequently attacks the shepherd's flocks" (91). This article by James Pierce was in fact published in 1823, but it is not discussed or referenced in the notebooks Darwin kept when moving to selection or when first articulating his theory. Darwin's was a genuine prediction in the sense specified above.

Moving on from the central mechanism, there were many other areas where the theory of the *Origin* is predictive in some sense or another. Not to belabor the point, I will mention only biogeographical distribution—that topic which was so crucial in Darwin's becoming an evolutionist in the first place—and I will give but one of the most successful predictions of the *Origin*, namely, that one would expect to find strong similarities between organisms on different islands of a group and those on the nearest major stretch of mainland, and less strong similarities between organisms on two geographically distant island groups, despite their having a similar habitat. Why, for instance, do the inhabitants of the Galapagos in the Pacific look so much like organisms from South America, whereas the inhabitants of the Cape Verde Islands in the Atlantic look so much like organisms from Africa, when the islands themselves have the same kind of volcanic habitat (398–399)?

> I believe this grand fact can receive no sort of explanation on the ordinary view of independent creation; whereas on the view here maintained, it is obvious that the Galapagos Islands would be likely to receive colonists whether by occasional means of transport or by formerly continuous land, from America; and the Cape de Verde Islands from Africa; and that such colonists would be liable to modifications—the principle of inheritance still betraying their original birthplace.

Move next to coherence and consistency. Darwin's theory is not obviously incoherent, in the sense of having blatant contradictions at its heart—anything but, in fact. Yet, there are some places where coherence did not come easily. Take the level at which natural selection operates: the underlying assumption throughout the *Origin* is that it is every individual against every other. "Hence, as more individuals are produced than can possibly survive, there must in every case be a struggle for existence, either one individual with another of the same species, or with the individuals of distinct species, or with the physical conditions of life" (63). This emphasis on the individual is reinforced by the secondary mechanism of sexual selection, where again we have intraspecific competition, with males competing for mates and females choosing those males that they like best.

However, internal coherence problems arose for Darwin when he attempted the consistent application of this individual perspective. Take the widespread sterility found in workers in the social insects. Given the emphasis on the individual, how could one explain the evolution of these nonreproductive castes? Eventually Darwin worked out a solution; but he had to compromise somewhat, arguing that here it is legitimate to treat the whole group as a kind of supraorganism (238):

> A slight modification of structure, or instinct, correlated with the sterile condition of certain members of the community, has been advantageous to the community: consequently the fertile males and females of the same community flourished, and transmitted to their fertile offspring a tendency to produce sterile members having the same modification. And I believe that this process has been repeated, until that prodigious amount of difference between the fertile and sterile females of the same species has been produced, which we see in many social insects.

Questions of heredity gave even more trouble with respect to coherence. Darwin's theory of natural selection, obviously, demanded a mechanism of heredity, to pass along traits from generation to generation. To this end, in the 1860s Darwin formulated his theory of "pangenesis," in which little gemmules, given off by all of the body parts, circulate around the body and eventually collect in the sex organs, from where they were ready to start the next generation (Darwin 1868; Geison 1969). But nobody, not even Darwin really, had much confidence in this theory of inheritance (Vorzimmer 1970). It was always ad hoc, never truly meshing with what was known already about organisms.

Its first problem, as friends pointed out, was that it ignored cell theory—an omission which invoked hasty and extensive revision. Then, as Darwin's cousin Francis Galton pointed out, Darwin gave no clue as to the medium through which the gemmules are supposedly transported. The blood will not do, as Galton showed through transfusion experiments. Somewhat testily, Darwin responded that he had never specified the blood as the key medium (letter to *Nature*, April 27, 1871)—which is true, but if not the blood, then what? Pangenesis thus faded because it did not cohere with the rest of biology.

Move on now to external coherence or consistency, where the question is that of the consistency of Darwin's theory with the science of his own day, not of ours. Very troublesome was geology. A process like natural selection, even if it is not always quite as leisurely as some have supposed, obviously demands masses of time. Darwin himself, in the first edition of the *Origin*, calculating from the supposed rate at which the (Sussex) Weald is being denuded, gave an estimate of 300 million years from the first mammals. This was savagely attacked by geologists (Phillips 1860), and the estimate was dropped from later editions, but the need for lots of time was still there.

A need made the more pressing as the physicists took over attack (Burchfield 1975). William Thomson (later, Lord Kelvin) calculated, primarily on the basis of the earth's cooling, that the earth might be as young as 25 million years—far too short a time for selection alone to operate (Thomson 1869). Darwin's theory was apparently inconsistent with physics, the leader of all the sciences—a matter made no easier by the fact that Thomson's brightest research student was none other than George Darwin, Charles and Emma Darwin's own son, who kept conveying back to his father all of the latest findings and calculations (Burchfield 1974)!

Darwin—and, for that matter, his fellow evolutionists—strove mightily to restore consistency between evolutionism and the physics of the day. Darwin himself, straightforwardly, began to rely more and more heavily on Lamarckian mechanisms such as use and disuse and the inheritance of acquired characters to speed up the evolutionary process. According to the Lamarckian theory of use and disuse, organisms change their characteristics by using (or failing to use) them: the giraffe's neck gets longer as it stretches for the highest branches, to take a seemingly obvious case. These newly acquired characteristics increase the organism's chances of survival and reproduction and are passed along to offspring. Alfred Russel Wallace (1870) more ingeniously adopted a sub-hypothesis about the great intensity of the Ice Ages, which he thought would much increase selection pressures and so accelerate evolution. The only one not to do anything was Darwin's great champion, the morphologist and paleontologist Thomas Henry Huxley (1869), who declared that his position was consistent with anything that the physicists were prepared to allow.

What gives this story an ironic edge is the fact that it was the physicists who were wrong—gloriously wrong! We now know of radioactive decay, its heat-producing effects, and the consequent slowing down of the earth's cooling. The 4–5 billion years since the formation of the earth has been quite enough time for evolution, even an evolution fueled primarily by natural selection. Physics and biology are again in harmony. But this happy relationship only underlines the key impor- tance of the value of external consistency, and whatever we may think today about the rights and wrongs of the debate in Darwin's time, he and his supporters had good reason to feel tense on this score.

We come now to one of the most significant epistemic values of them all: unificatory power. Here we can speak confidently. Above all else, Darwin's theory exhibits this value, unifying into one whole all of the hitherto disparate areas of biology: paleontology, biogeography, behavior, embryology, systematics, morphology, and more. One starts with a central causal core: primarily the fact of evolution through natural selection brought on by the struggle for existence among organisms with different characteristics; to this is added subsidiary mechanisms like sexual selection, as well as those dealing with specific issues like the principle of divergence to explain speciation, and also speculations about heredity. Then this formulation is applied throughout the range of biology. Thus (456):

> I have attempted to show, that the subordination of group to group in all organisms throughout all time; that the nature of the relationship, by which all living and extinct beings are united by complex, radiating, and circuitous lines of affinities into one grand system; the rules followed and the difficulties encountered by naturalists in their classifications; the values set upon characters, if constant and prevalent, whether of high vital importance, or of the most trifling importance, or, as in rudimentary organs, of no importance; the wide opposition in value between analogical or adaptive characters, and characters of true affinity; and other such rules;—all naturally follow on the view of the common parentage of those forms which are considered by naturalists as allied, to- gether with their modification through natural selection, with its contingencies of extinction and divergence of character.

And so on and so forth for everything else. Unification is the name of the game (Ruse 1979; see figure).

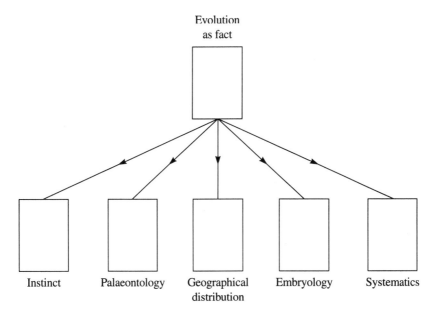

Evolution
as fact

Instinct Palaeontology Geographical Embryology Systematics
distribution

The unificatory nature of Darwin's theory.

Next we have fertility. Taken at its broadest, it is part of what one modern philosopher of science characterized as a "progressive research programme," a theoretical framework that keeps throwing up new problems and challenges, together with the resources to deal with them (Lakatos 1970). Without prejudging the issue of how people did take up the ideas of the *Origin,* one can surely say that Darwin's theory carried within it much potential for masses of new work in altogether new directions. Darwin himself shows this, primarily in the way that he refers to so many different areas of biology; and if he did not always give definitive answers, he was forever scratching away at problems.

More than this, there are—there were—ways in which the theory could be extended into new fields toward new problems. Consider the work of a close friend of Wallace, Henry Walter Bates (1862), as he looked mimicry in Amazonian butterflies. He was able to show that this is a problem Darwinism sets and is able to solve, triumphantly. The models (those being mimicked) are poisonous; the mimics are not. However, thanks to similarity of appearance, the mimics are able to piggyback to safety, because the major predators, birds, have learnt to avoid all like-colored prey. Moreover, pushing his work to the limit, Bates made predictions about the points at which mimicry might break down or prove less than definitive, and these predictions bore out completely.

Simplicity is in the eye of the beholder, somewhat. It is not easy to say whether in itself the *Origin* was simple or elegant, unless (like Whewell) one reduces the notion to something else, like unificatory power (in which case it is certainly elegantly simple). With respect to this epistemic value, one really needs to go back and see how Darwin and his contemporaries regarded the theory. So for now, let us conclude the direct analysis of the epistemic values. Although it was certainly not fully effective in every respect, the theory of the *Origin* was, to borrow a metaphor from another science, a quantum jump above anything offered by grandfather Erasmus. But what about the nonepistemic or cultural values influencing Darwin's work?

Cultural Values in Darwin's Science

Start with religion. As we have seen, the young Darwin moved from Christianity to deism, and evolution was for him, as for his grandfather, a confirmation of his religious position rather than an anomaly. This was the philosophy of the *Origin*: "Authors of the highest eminence seem to be fully satisfied with the view that each species has been independently created. To my mind it accords better with what we know of the laws impressed on matter by the Creator, that the production and extinction of the past and present inhabitants of the world should have been due to secondary causes, like those determining the birth and death of the individual" (488). Later in life, particularly under the influence of Huxley, Darwin's beliefs faded into agnosticism. Even then, however, he did not go through the *Origin* systematically removing references to God. Although the sixth edition of the *Origin* (1872) was 75 percent rewritten and augmented from the first edition (1859), the just-quoted passage was left intact. There are nine references to the Creator in the sixth edition.

What about natural theology? Again, we have seen a religious influence. As a young man, Darwin read Paley's *Natural Theology*, and he absorbed and accepted absolutely the message about design. Organisms seem as if they were planned and made by an artificer. The distributions of the Galapagos reptiles and birds made no sense, without something to account for the nonrandom pattern. The beak of the finch made no sense, without something to account for the perfection of its adaptation. Evolution in itself spoke to the first, and natural selection—a designlike producer—spoke to the second. Right through the production of the *Origin*, Darwin shows appreciation for God's actions as a law-bound Creator. God as teleologist.

Later, as Darwin started to downplay natural selection, in some respects he loosened his ties to a ubiquitous intricately designlike organic world. But to the end, for all of the personal agnosticism, the design-seeing cultural influence persisted—in a major way. For example: "It will be admitted that the flowers of orchids present a multitude of curious structures, which a few years ago would have been considered as mere morphological differences without any special function; but they are now known to be of the highest importance for the fertilization of the species through the aid of insects, and have probably been gained through natural selection" (Darwin 1959, 235). This passage was added to the sixth edition of 1872; in all, in this edition there are 28 references to "adaptation" and 31 references to "function."

After religion, what of secular philosophies? It is often thought that, unlike his grandfather and every other evolutionist before the *Origin*, Darwin had foresworn progress. After all, natural selection is a relativistic process that does not in itself favor one type or species over another. However, to conclude this is to see Darwin out of context. It is simply not right to think he had abandoned progress (Richards 1992). We know that as an individual Darwin was always a social progressive, although in the first edition of the *Origin*, probably intentionally to separate himself off from the pseudo-science of the past, Darwin was restrained on the subject of biological progress. But his progressionism is there in the metaphors and language, most particularly in the flowery prose which ends the volume. By the time of the third edition, only two years later, the progressionism is even more overt (Darwin 1959, 222):

> If we look at the differentiation and specialisation of the several organs of each being when adult (and this will include the ad-

vancement of the brain for intellectual purposes) as the best standard of highness of organisation, natural selection clearly leads
towards highness; for all physiologists admit that the specialisation of organs, inasmuch as they perform in this state their func-
tions better, is an advantage to each being; and hence the
accumulation of variations tending towards specialisation is within
the scope of natural selection.

Progress is there and it is valued—a point which becomes yet
clearer when Darwin turned to *Homo sapiens*. Although there is little on
this in the *Origin*, in *The Descent of Man* Darwin brought in all of the
cultural values of his sex, race, and class. Not only do we learn that men
are strong and brave and brainy, whereas women are kind and gentle
and sensitive; that whites are intelligent and hard-working whereas
blacks are stupid and lazy; but that, on the whole, capitalism is no bad
thing (Darwin 1871, 1, 169):

> In all civilized countries man accumulates property and bequeaths
> it to his children. So that the children in the same country do not
> by any means start fair in the race for success. But this is far from
> an unmixed evil; for without the accumulation of capital the arts
> could not progress; and it is chiefly through their power that the
> civilised races have extended, and are now everywhere extending,
> their range, so as to take the place of the lower races. Nor does the
> moderate accumulation of wealth interfere with the process of
> selection. When a poor man becomes rich, his children enter
> trades or professions in which there is struggle enough, so that the
> able in body and mind succeed best. The presence of a body of
> well-instructed men, who have not to labour for their daily bread,
> is important to a degree which cannot be over-estimated; as all
> high intellectual work is carried on by them, and on such work
> material progress of all kinds mainly depends, not to mention
> other and higher advantages.

Culture was clearly the major motive force at this point. With this
very Victorian notion of progress—if ever there was a self-serving argument, it was this passage by the grandson of Josiah Wedgwood—we
have an entry to many other values to be found in Darwin's work. But
rather than just attempting an exhaustive catalogue, let me conclude this

part of the discussion by noting how one of my own predictions is already starting to bear fruit. I forecast that we would start to see nonepistemic cultural factors—values even—coming to bear on epistemic ends, as these cultural entities support the ever-greater satisfaction of such aims as predictive success and unificatory power. Even with Erasmus Darwin there were hints of this happening, and now with Charles Darwin the moves become open.

Deism, to take the prime example, is nonepistemic, cultural. It is hard to deny that values are involved in the minds of the two Darwins. Certainly, they have a very positive picture of their God. But inasmuch as this deistic commitment leads to a belief in a law-bound world, it is surely preparing the way for the satisfaction of epistemic values like predictive accuracy. The nonepistemic promotes the epistemic. Although, bearing out a second prediction, do note how this is happening. Charles Darwin's deistic God is not part of his science. That is the whole point. God has been taken out of the science. Rather He (or It) is functioning as a metavalue. By this I mean (as does the mathematician when talking about metamathematics) something which is about the subject rather than within it. Darwin's God is setting or promoting the conditions for good science. He (It) is not within the good science. Hence, there is no cultural value within the science, challenging the autonomy of the epistemic values, but there is a cultural value *without*—outside—the science, promoting the internal epistemic values.

None of this makes impossible a continued use of the epistemic/nonepistemic dichotomy. But it does start to prepare the way for the claim that the objective/subjective division may be more complex and less clean-cut than extremists allow.

Charles Darwin's Science in His Own Time

Thus far, I have been considering Darwin's theory from our perspective. We have seen a huge jump in epistemic power from the theorizing or speculations of his grandfather. Yet at another level, nonepistemic factors (including values) continue to ride high. One would hardly say that the only reasons for accepting Charles's theory were nonepistemic ones, which was more or less the case for Erasmus. But one could well

imagine that someone unsympathetic to Charles's cultural values—someone indifferent to natural theology and hating modern industrialism—might have difficulty with his theory.

The time has come to leave the hypothetical and turn back the
clock. How in fact was Charles Darwin's theory regarded by his contemporaries—or by himself, for that matter? Even if we disregard social
factors—by the time the *Origin* was published Darwin was a distinguished member of the scientific establishment—one would expect a
very different response to the *Origin* than to any of his grandfather's
writings. The epistemic values we have been discussing were known and
valued in 1859, and Darwin had made major efforts to see that they were
respected and satisfied (Ruse 1975). The nonepistemic values might not
have been appreciated by everyone—Thomas Carlyle in Britain and
Charles Sanders Peirce in America spring to mind as people who
loathed much of the traditional value ethos—but, taken as a whole,
Charles Darwin's values were much more in tune with society generally
than his grandfather's had been. In major respects, of course, this was
because the Darwins had stood firm and society had changed around
them. Many of Charles Darwin's cultural values—religion, progress,
industrialism—were precisely those of Erasmus Darwin.

Overall, therefore, our prediction would be that Darwin's work
would definitely be elevated from the depths of pseudo-science. But
would we expect to find it regarded as science of the front rank? Would
it be science that might give us definitive answers about the questions
of objectivity and subjectivity? My suspicion is that it would not, because although Darwin went far along the epistemic route, he clearly did
not reach all of his end points. Even if we forget such issues as the
(then-apparent) inconsistency with physics over the age of the earth,
Darwin's predictions are in no sense quantifiable, for instance. Often,
as with the wolf example, we get more in the nature of promissory notes
than hard-line evidence. And the same is true elsewhere. Right through
to predictive fertility, we have something for the future, more than
something actually achieved in 1859 or the years shortly thereafter.

The forecast would surely be that, judged by epistemic criteria and
recognizing that opposition based on cultural or nonepistemic values
would now no longer be a crucial barrier, the fact of evolution might

well be accepted—the unificatory power would speak to this—but that the proposed mechanisms or causes might be received with less enthusiasm. And this is how it proved. For all that tradition has it that Darwin's evolutionism faced monstrous opposition, the truth is that evolution per se (evolution as fact) became orthodoxy almost overnight (Ellegard 1958). Like the emperor's new clothes, once Darwin had spoken—wrapping his ideas in such a socially acceptable form—most people were happy to slip over and accept descent with modification. This applies to Believers as well as others. It is true that some of the old guard, like the Swiss-American ichthyologist Louis Agassiz, could never accept evolution in any guise. But all of his students did (Hull 1973). I am ignoring those who never could, then or now, accept evolution because of religious reasons. However, although they continue to make a big noise, this group was always a minority (Numbers 1992).

People were a lot less enthusiastic about natural selection, however (Bowler 1988). No one denied its existence or its force. Indeed, to answer positively the unanswered question about simplicity, we find Huxley chiding himself for missing the idea of selection: "How exceedingly stupid not to have thought of that!" (Huxley 1900, 1, 170). But there was a feeling that natural selection needed major supplement to get real results. Hence, some made Lamarckism the chief causal mechanism. Others preferred saltationism, the postulation of instantaneous jumps from one species to another. Yet a third group endorsed orthogenesis, the notion that life forces somehow push organisms up the chain of life. And some, like the American botanist Asa Gray (1876), who was Darwin's champion in the New World, nevertheless thought that a divine guiding element somehow entered into each new successful variation—on which natural selection could then work, deprived of any creative role but weeding out the losers and the inadequate.

It was not just the epistemic inadequacy of the *Origin* that was at work here: people like Gray were open in their religious motivations. But the epistemic shortcomings of Darwin's evolutionism was a major factor, perhaps more so, or in more direct ways, than it would be today. In our modern age of quarks, black holes, cyberspace, and other intuitively strange entities, raw empiricism—the insistence that one have a physical, hands-on picture of reality—is generally not rated that highly.

The French particularly have been scathing on the subject. Of an account of electricity cast in terms of pulleys and cogs and cords, Pierre Duhem ([1906] 1954, 71) remarked sarcastically: "We thought we were entering the tranquil and neatly ordered abode of reason, but we find ourselves in a factory." At the time of the *Origin*, however, the call to provide directly sensed evidence was heard repeatedly, especially in the circles of British science. "I never satisfy myself until I can make a mechanical model of a thing" (Thomson 1884, 131). And this was the problem: no one sees evolution actually occurring before their eyes.

Or was it a problem? Here we encounter a paradox. Not everyone sympathetic toward evolution wanted to ignore the empiricist's call for direct evidence. This applies especially to those who might have their own personal reasons—nonepistemic reasons—for taking the empiricism demand seriously. One such person was Huxley (Desmond 1997). In the second half of the nineteenth century he and his friends were pushing hard to make a space for science in British society, especially including British education. Huxley was open in wanting to replace study of the classics with his own speciality, morphology; and to this end he was forever trumpeting the moral virtues of individual empirical experience. There is an intentionally biblical echo to his most famous dictum: "Sit down before fact as a little child, be prepared to give up every preconceived notion, follow humbly wherever and to whatever abysses nature leads, or you shall learn nothing" (Huxley 1900, 1, 219). Since morphology, unlike Bates's field of insect camouflage, is not a subject for which natural selection offers much of value as a working tool—the morphologist spends most of the time with dead organisms, trying to discern homologies between one specimen and an-other—Huxley therefore had the liberty and luxury to criticize Darwin's mechanism on the grounds that, inasmuch as no one had ever used selection to produce a new species, it failed the test of empiricism. Selection may have been obviously true; it was not obviously effective.

Darwin had tried to anticipate this sort of criticism by making much of the analogy of artificial selection, a humanly sensed equivalent to the unseen supposed mechanism of natural selection. But when the attacks came, eventually he had to respond that empiricism is not the only or even the essential criterion of good science. "In scientific investigations it is permitted to invent any hypothesis, and if it explains various large and independent classes of facts it rises to the rank of well-grounded theory. The undulations of the ether and even its existence are hypothetical, yet every one now admits the undulatory theory of light." Likewise for selection: "If the principle of natural selection does explain . . . large bodies of facts, it ought to be received" (Darwin 1868, 1, 8–9). But essentially Darwin could not win this battle, not at this time at least. His critics had their own nonepistemic ends to please.

These nonepistemic ends of Darwin's readers were analogous to the deistic ends of Darwin himself: they were directed toward making science *more* epistemic. Which brings me to a major factor in the then-received status of the theory of evolution through natural selection, one which will take on an increasing role in our story: the belief and directive that the best science, the science of the professional, be free of nonepistemic values; that science be objective. Not that it necessarily *is* objective but rather that it is cherished inasmuch as scientists *think* that it is objective. We are talking here therefore of a metavalue, one that is *about* science, rather than *within* science.

This metavalue, objectivity, did not arise only in the context of the *Origin*. Earlier in the century in France, the metavalue of cherishing objectivity had been a constant refrain of Cuvier (Ruse 1996). He was forever criticizing evolutionists and German morphologists (the *Naturphilosophen*) and natural historians (especially the followers of the eighteenth-century Comte de Buffon), on the grounds that illicitly they brought cultural values into their science. Given the clear cultural elements in much that he passed as science, it is tempting to accuse Cuvier of rank hypocrisy. However, in mitigation, one should note that a major personal factor was driving the great anatomist. He was a modestly born Protestant in a country which (especially in the post-Napoleonic era) was increasingly conservatively Catholic. To make his way, therefore, he had to convince people that his field, science, was the one area above all where personal value systems, whatever they might be, are irrelevant. Cuvier's proto-Popperian position was a foundation plank (Outram 1984).

As it was also for the biologists of Darwin's day. Huxley simply had to convince people that science is free of cultural values: that his science was beyond or outside culture. Else he would have had absolutely no success whatsoever in persuading his countrymen to allow science any role in any form in educational curricula. As he and his friends led family-centered lives of the strictest Victorian rectitude, so also he and his friends claimed to work on matters of the strictest value neutrality. Sex, religion, race, class were (supposedly) stopped at the classroom door, just as George Eliot—whom Huxley knew and admired—was stopped at the Huxleys' front door, lest her sinful nature (she lived with a man to whom she was not married) contaminate the angels at the hearth. Students in Huxley's anatomy course spent hours dissecting a cat, not excluding its genitalia, and listened to several lectures on the features distinguishing Africans from Europeans, but perish the thought that these studies would tell them anything about themselves.

Huxley and others were much attracted to the morphological musings of the great German evolutionist Ernst Haeckel. But until Haeckel showed himself willing to tone down all of the radical philosophical proposals strewn through his masterwork *Generale Morphologie*, there was no way the Huxleyites were going to endorse the translation of that sort of stuff into English (Bowler 1996). As had been the case for Cuvier, it was an article of the strictest faith that science and values are friends and not partners.

Evolution as Religion

Darwin himself knew the rules. As a member of the professional scientific elite, he was aware that science has a role—one it would like to increase—in society, and part of that role and status is that, in its professional, official form, it be nonepistemic-value-free. Science is supposed to be different from religion and philosophy and the like: it is supposed to be beyond or without culture, and objective. As we would expect, Darwin frequently shows himself uncomfortable with overt nonepistemic value claims in science—professional, would-be mature science, that is. Yet, comfortable or not, such values did not all vanish, even from the *Origin*. In the case of progress, we have seen how, if anything, its role was increased and intensified. So apart from epistemic failings, here again we have reason why we should not expect to see Darwin's science raised to full maturity.

Which brings us to one of the most fascinating items in the whole history of evolutionism. Whatever Darwin's own aspirations, his friends and supporters—Huxley particularly—had very little interest in seeing evolution raised to the level of a fully mature, objective, culture-free science (Ruse 1996). Huxley wanted a role in Victorian society for culturally neutral science—his own morphology for general education and (especially through his students Michael Foster and H. N. Martin) physiology for the medical profession. But this does not mean that he had no nonepistemic values of his own or no desire to promulgate them. He most certainly did—progress, meritocracy, a functioning well-run state, secular education, improved medical care, the diminution of inherited privilege, and more—and he needed a medium through which to effect his ends. Conventional Christianity was not going to serve this purpose; even if there had not been the matter of belief, bishops and others already occupied the favored places. Huxley and friends had therefore to set about creating their own favored secular religion or philosophy. And evolution—which, remember, was considered to be epistemically below par anyway—was the perfect vehicle.

Hence, what we find is that no attempt was made to expel the culturally value-laden from evolutionary biology. If anything, it was welcomed and made more explicit. Huxley's first and most influential work on evolution was a popular book on the history and nature of our own species *(Man's Place in Nature)*. It is true that, by the end of the century, we find a kind of second-rate evolutionary morphology—tracing paths, relying on a vanishingly weak connection between individual and group development (the so-called biogenetic law that phylogeny recapitulates ontology, which is to say that the entire evolutionary development of a given species is repeated in the embryologic development of each individual member of that species) and offering as many solutions as there were people writing on the topics (Nyhart 1995). However, essentially Huxley and his followers had little interest in anything more. Evolution—naturalistic, progressivist, all-encompassing—was the perfect ideology, the made-for-the-purpose secular religion. There was no drive to go further, to a fully mature science.

And in line with the strategy of keeping culture out of the classroom, evolution was simply not a topic to be taught in schools or universities—a fact noted by one of Huxley's puzzled students (a Jesuit): "One day when I was talking to him, our conversation turned upon evolution. 'There is one thing about you I cannot understand,' I said, 'and I should like a word in explanation. For several months now I have been attending your course, and I have never heard you mention evolution, while in your public lectures everywhere you openly proclaim yourself an evolutionist'" (Father Hahn, quoted in Huxley 1900, 2:405). Darwin's greatest supporter had a very definite program of his own to follow.

Indeed, having done the spadework of bringing people over to the fact of evolution, Darwin then found that the world rather turned away from him. It wanted someone who could give it full-blooded moral messages, as one expects of any truly functioning religion. It wanted a prophet of evolution, a role happily assumed by Herbert Spencer (Richards 1987). In a series of works of Wagnerian scope, his "synthetic philosophy," Spencer unrolled an eclectically imagined, all-encompassing picture, where everything—inorganic, organic, human, cultural—is seen as part of an upward metaphysical movement, now pausing in a state of equilibrium but then gaining strength for further progress.

Totally ignored were worries of the physicists that the newly discovered Second Law of Thermodynamics suggests that the whole of creation is on an inevitable downward path to static heat death. Rather, from early maximal fertility (think of the number of offspring of the herring) the Spencerian world progresses to minimal fertility (epitomized by the upper-class English), as life's vital resources are used in the making of intelligence rather than sperm cells and the initially homogeneous is transformed through Lamarckian-ingrained effort into the thoroughly heterogeneous (Spencer 1857, 1862, 1864, 1892).

With this eschatological fantasy came the exhortation to maintain and succor the forces of evolutionary destiny. It is often thought that Spencer preached an extreme *laissez faire* social Darwinism: the weakest to the wall and may the strong survive. This is but part of his system, and a minor part at best. Much more significant is the progressive drive to make here on earth what the Christian offers only providentially: the Kingdom of Heaven. And indeed, just as we find Christians divided bitterly over the correct means to the end—the Quaker who preaches pacifism matched by the army padre who preaches force—we find evolutionists divided, sometimes bitterly, over their ends (Crook 1994; Pittenger 1993; Mitman 1992; Russett 1989).

In America especially, where social evolutionism took deep root and where Spencer's works far outsold those of Darwin, we find that his synthetic philosophy appealed to all shades of opinion, and often to the most divided of rivals. Socialists saw in Spencer the very optimism for the future that they found in Marx; businessmen, however, found the libertarianism of Spencer to be an excuse for union-busting. Pacifists found in Spencer support for world peace, inasmuch as militarism was taken to be a bar to free trade; the warlike, however, found in Spencer reason to prepare for coming struggles for existence. Feminists liked the idea of progress toward full rights and education for women; male chauvinists found in Spencer support for the view that only the rightful natural-born leaders (men) should have the vote.

Darwin's thinking and that of Spencer overlap—they both believed in organic as well as social progress—but in many respects the two visions are very far apart. For all his increased use of Lamarckian mechanisms in later editions of the *Origin*, Darwin always considered these secondary to natural selection. For Spencer, notwithstanding the fact that he hit independently on selection and contributed the phrase "survival of the fittest" to the language, the inheritance of acquired characteristics always outweighed selection in his thinking.

Even more crucial is the question of equilibrium. This notion, as used in a biological context, has its roots in the ancient doctrine of a balance of nature—God has so organized the world that predators and prey balance each other out so that there is always food for the one and space for the other (Egerton 1973). Darwin essentially (and properly) saw this belief, which was held by the natural theologians of his own day, as being superseded by evolution through selection—a move from the essentially static to the essentially dynamic. In the first edition of the *Origin*, the balance of nature gets minimal treatment, if that. Spencer (whose thinking was rooted deep in the nonconformist Christianity of the British Midlands) retained the balance but adapted it by dressing it up in the language of the physics of the day, giving it a dynamic twist so that it could play a central role in his progressionist world picture: "Throughout Evolution of all kinds there is a continual approximation to, and more or less complete maintenance of, this moving equilibrium" (Spencer 1862 [1912], 451).

Charles Darwin could not stop or delay the rush to make of evolution a modern secular substitute for the older religions. Indeed, in some ways one senses that, toward the end of his life, he sank into acquiescence. He had status and honor and respect—remember the Abbey. But he had and was to have no full-blooded science. So, going with the flow, Darwin became more and more overt about his own social values. Not only did he stiffen up the discussion on progress in the *Origin*, but he even added a reference to Spencerian equilibrium (which became more commonly known as "dynamic equilibrium"). And Darwin paid more attention to that supremely cultural-value-attracting animal, man. In some respects, *The Descent of Man* is far more a work for the popular domain than is the *Origin*. But this is what one expects and wants from a popular science, a religion substitute. Lots of stuff on origins and ethics and social customs. Just like the Bible. Hence, with Huxley and Spencer and all of the others, this is what Charles Darwin fell to providing. Aims for tough-minded mature science were put on the back burner.

Into the Twentieth Century

In the first hundred years of its history, evolution rose up from its origins in the lowest depths of pseudo-science. Epistemic values started to play a role—a significant role—in the theory's construction and defense. But culture rode high. Indeed, there was a thorough and systematic mixing of the epistemic and the nonepistemic. For all we have started to tease apart these connections, we still cannot yet draw firm conclusions. Evolutionary thought is more epistemically rigorous than it ever was; yet at all levels it is thoroughly impregnated with culture. It is no longer mere quasi- or pseudo-science; but at the end of the nineteenth century, evolution is considered by no one to be science of the first order, as the physical sciences were. Evolutionary biology was the science of the public domain. That was where it found its support and that was the audience to which it catered. It was a kind of secular religion.

What we must now see is how things were to change in our own century.

4

✑

JULIAN HUXLEY

Religion without Revelation

May the eighth, 1900. The biologist William Bateson was on the train, traveling to deliver a lecture on problems of heredity at the Royal Horticultural Society (Bateson 1928). His reading material for the journey included work by the hitherto-unknown Gregor Mendel, the mid-European monk who, even at the time of Darwin, had discerned the essential rules of heredity. Bateson was so delighted that he scrapped his original talk and gave instead an enthusiastic lecture on Mendelism. One wonders if his audience knew that they were hearing about the most important conceptual advance in the post-Darwinian history of evolutionism: the basis of an adequate theory of heritable transmission or "genetics."

Things moved quickly. Thanks particularly to Thomas Hunt Morgan and his assistants at Columbia University in New York City, the second decade of this century saw a unifying synthesis of Mendel's findings with new insights into the physical nature of the cell: the "classical theory of the gene" (Allen 1978a,b; see box). Many have told the story of how this new genetics was integrated with Darwinian selection, to create neo-Darwinism or the "synthetic theory of evolution," so it is not necessary for us to linger long here. But in order to lead up to the main subject of this chapter, let us set the scene by mentioning briefly the ways in which the integration occurred.

Mendelian Genetics

Organisms are composed of *cells*, within the centers (the *nuclei*) of which are stringlike entities, the *chromosomes*. *Genes*, the units of heredity, are found along the chromosomes, and the entire set of genes within a cell (the *genotype*) is the same (with some exceptions) throughout the body of each individual. In addition to being the units of heredity, the genes are also the units of function, in a sense the blueprint for building the physical body, the *phenotype*.

In sexual organisms, with some few exceptions, the chromosomes are paired, and each matching point is known as a *locus* (plural, *loci*). The different versions of a gene which can occupy any specified locus of the members of a species are called *alleles* (formerly, *allelomorphs*). If two alleles at some locus in an individual are identical, then with respect to that locus the individual is said to be a *homozygote*; if the alleles are different, the individual is a *heterozygote*. The genes are very stable, but occasionally they change, forming *mutations*; genetic mutations cause new variations in the phenotype.

The key Mendelian claim, the *law of segregation*, is that copies of the genes are passed entire to offspring; they are unchanged (with the exception of mutations) and are not in any sense blended with other genes. In sexual organisms, each offspring receives half of its genes from one parent and the other half from the other parent. At any particular locus on the chromosomes of the offspring, one of its genes comes from one parent's corresponding locus and the other from the other parent's corresponding locus. It is equiprobable as to which of each parent's two genes at each locus will be transmitted.

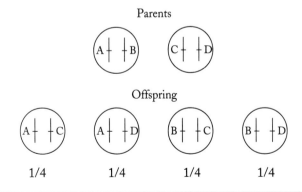

The Genetics of Populations

Although the early Mendelians were bitter personal opponents of the Darwinian selectionists (the "biometricians"), people interested in evolution's mechanisms realized very quickly that a theory which allows the units of heredity to be passed on untouched from generation to generation and which sees "mutations" (new versions of the units of heredity) to be infrequent, small, and random (in the sense of not appearing to respond to a "need") is the perfect complement to natural selection. It took some years, however, for these ideas to be combined formally, a task which was accomplished around 1930 by several mathematically minded biologists, notably Ronald A. Fisher and J. B. S. Haldane in Britain and Sewall Wright in America (Provine 1971).

Notwithstanding their mathematical genius, one should be wary of overestimating the achievements of these theoreticians. We have seen how the status of evolutionary theory in the years after the *Origin* was ambiguous, to say the least. It was no longer pseudo-science, which no respectable person would acknowledge, particularly no respectable professional scientist. Indeed, virtually all people became evolutionists. But for many, especially the scientists, evolution became a kind of metaphysical background that conveyed a cultural message—primarily the message of progress but also the rightful role of the sexes and the significance of various political doctrines—and there was little inclination to disturb the status of this very convenient vehicle.

Even those who wanted to make something more of evolution, and who brought it to the foreground of their professional activities, generally spent their days in a fantasy land of their own making, building highly speculative pictures of life's past. Since the usual places where such people found homes and support were museums—institutions whose primary purpose was to provide amusement and instruction to the public—it is little wonder that their work frequently sacrificed the epistemic for the cultural, as generations of schoolchildren were fed moral and social messages from the past, sandwiched with important information about the significance of personal hygiene and deference to superiors (Rainger 1991).

The mathematical evolutionists—"population geneticists"—did

not themselves bring about a complete change in the status of evolutionary thought. Although their work could be made formally compatible
with the way they saw the processes of evolution actually working, there
were major differences separating these men, Fisher and Wright in particular (Hodge 1992). Disputes were as bitter as between selectionists, sal-
tationists, and orthogeneticists. Nor, for all the mathematics, did they
drive out culture. Indeed, culture was the very cause of many of the differences, and its elimination was furthest from their minds. Fisher's population genetics was part of a grand metaphysical scheme as wonderful as
anything seen in the nineteenth century, and Wright's was little different
in this respect. They could both have given Herbert Spencer a good run
for his money—except that they (Wright especially) were running with
Spencer rather than against him.

Fisher's passions were eugenics and Christianity (Box 1978). He
believed that God created the world in a progressive fashion, working
up to humankind. Selection, which is most effective in large populations, is always moving organisms up to their peak of adaptive fitness.
Even though environmental degradation and so forth is forever frustrating this end, the overall effect is one of ever-higher forms of life.

Unfortunately in the human case, thought Fisher, improvement in
biology leads to improvement in cultural forms. These lead to greater
wealth and social status, at which point evolution backfires, as those at
the top start to restrict their breeding. The poor and ignorant, mean-
while, continue to have very large families. Biological degeneration sets
in, and this can be prevented only through whole-scale eugenic practices; through such things as tax rebates, the state must encourage the
upper classes (the repositories of the better genes) to reproduce more
than they would otherwise. Not to act in this way is to turn from our
Christian duty.

One cannot overemphasize the extent to which these views informed Fisher's thinking about the evolutionary process. The very idea
that selection moves organisms slowly up to a peak of fitness, in a kind
of converse way to the decline caused by the Second Law of Thermodynamics (Fisher's analogy), was the key to his whole vision of the
evolutionary process—a vision which was not shared in any way by
Sewall Wright in America.

For Wright, the key to evolutionary change was the breaking of large populations into small subpopulations, genetic divergence within the latter, and then a violent shaking down as the subgroups rejoin (Wright 1931). Very significant for Wright was the fact that, if one has a subgroup sufficiently small and long-lasting, the purely contingent effects of mating and the random effects of Mendelian transmission can outweigh the force of selection. This essentially accidental and nondirected form of change has come to be known as *genetic drift*, or the Sewall Wright effect (Provine 1986; see box).

The facts of nature were not totally irrelevant for Wright, any more than they were for Fisher. For ten years, the American worked for the U.S. Department of Agriculture, and an intensive study of cattle breeding convinced him that the most effective methods required fragmentation, change within small groups, and then a return to the whole population. But ultimately, the philosophical and cultural factors were as significant for Wright as they were for Fisher. Behind Wright's theorizing lay the teaching of his Harvard professor of chemistry, L. J. Henderson (1913, 1917). Following the older man, Wright was an enthusiastic Spencerian, believing in particular that nature tends perpetually toward a state of moving or dynamic equilibrium: balance and homogeneity are forever being disturbed by external factors; these disruptions provoke forces trying to return the system to equilibrium; and in the overall process, heterogeneity increases and nature progresses upward to a better state.

Wright called his theory of evolution the "shifting balance theory," meaning that one got an uneasy or ever-moving balance between the forces for homogeneity and those for heterogeneity: division of populations and genetic drift break things up and create diversity, and then the subsequent rejoining of populations and the natural selection that now kicks in creates more uniformity. "The type of moving equilibrium to be expected, according to the present analysis . . . agrees well with the apparent course of evolution in the majority of cases" (Wright 1931, 154). Very important here was Wright's metaphor of an "adaptive landscape," which saw populations of organisms as sitting on the tops of adaptive peaks; then, thanks to disruptive forces, these organisms are driven down to the less-adapted valleys and so up the sides of new peaks (Wright 1932; see figure).

Population Genetics

Mendel's law of segregation applies to individuals but can be readily generalized to groups. Suppose for simplicity that one has just two alleles, A and a, at some locus in a large population and that the ratio of A to a is p to q (where by definition p + q = 1). Then the so-called *Hardy-Weinberg law* states that, if breeding is at random, the ratio p to q will stay the same and that, after the first generation, the distribution of genotypes will be fixed by a simple formula.

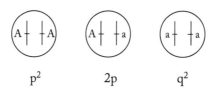

$$p^2 \qquad\qquad 2p \qquad\qquad q^2$$

This law implies that, all other things being equal, even if a gene is very rare, it (and its phenotypes) will persist indefinitely in the population. As Newtonian mechanics uses the First Law (bodies remain at rest or in uniform motion unless acted upon by a force) as a background assumption against which one can introduce agents of change, so population genetics uses the Hardy-Weinberg law as a background assumption against which one can introduce agents of change, for example mutation and natural selection. One can study the effects of such causes over time, confident that things will not just follow in a haphazard fashion.

The law holds only for large groups. In small groups, chance factors become significant. Even though selection may favor a particular allele, chance effects might swamp its virtues and drive it to extinction within the group (or, conversely, establish as universal an allele not favored by selection over its fellows). At any time, one might expect to find some variation not under the tight control of natural selection. This is the key to Sewall Wright's concept of *genetic drift*. Just how small groups have to be and how many generations are required for certain effects to become likely or probable can be demonstrated mathematically.

It was certainly not a necessary part of the picture, but Wright interpreted all of this in a progressionist fashion. Indeed, this was a major attraction of the metaphor. For all that the environment might change, and peaks might no longer be so very peakish, or new peaks might be lower than old peaks, Wright thought that the overall effect over time is upward. Indeed, we know that he had a metaphysical view

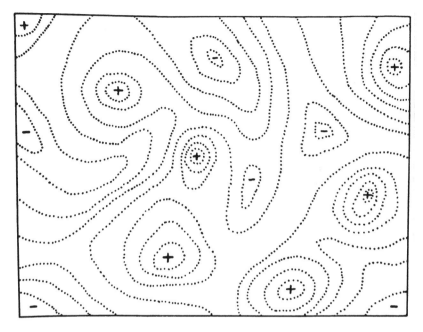

A bird's eye view of an adaptive landscape (from Wright 1932).

of "panpsychic monism," something which he got from the English philosopher W. K. Clifford (1879) via the philosopher-statistician Karl Pearson (1892). Wright believed that all of life is a combination of both mind and body and that there is a progressive scale whose summit is a universal mind/body, a kind of naturalistic, post-Hegelian Absolute. This belief was certainly not part of Wright's science—like his teacher, Henderson, he believed that one should keep one's science and one's metaphysics strictly separate—but such a belief was the framework into which Wright was happy to mold his science.

Just as one should be wary of assuming that the population geneticists single-handedly upgraded (or wanted to upgrade) evolutionary thought to the level of what we (or they) would consider fully mature science, so also should one take care in assessing the relationship between these men and those who came later and took up their thought. By and large the biologists of yesterday were not very mathematically inclined. Indeed, one can put matters more strongly and say that they tended to flee from figures. Hence, one does not find that other evolutionists simply took the formalisms of the population geneticists and clothed them with empirical findings. As far as the mathematics is concerned, it functioned more as an inspiration and encouragement that evolutionary thought could be put on a formal basis if one was so inclined. At least as important, the very existence of the formal work gave a weapon to turn on those doubters and critics who sneered that evolution deserves its second-rate status as a science for the museums.

By the 1930s, an increasing number of young biologists, ardent in their evolutionism, wanted such a weapon. They were growing frustrated with the low academic status of evolution—its exclusion before such subjects as genetics and experimental embryology and physiology—and were forming the resolve to upgrade evolution as a science (Cain 1992, 1994; Smocovitis 1992). These people are the main focus of this and the next chapter. I choose two representatives: for this chapter, Julian Huxley, an Englishman who was ultimately unsuccessful; and for the next chapter, Theodosius Dobzhansky, a Russian-born American who was ultimately successful.

Julian Sorell Huxley (1887–1975)

Julian Huxley was the oldest grandson of Thomas Henry Huxley, "Darwin's bulldog." Educated at Eton and Oxford, he was, like his grandfather, a biologist, a prolific writer, and, by the end of his life, a very public figure (Huxley 1970, 1973; Waters and van Helden 1992). However, unlike his grandfather, Julian Huxley disliked university life, and after short-term spells at Rice Institute (now Rice University) in Texas, back at Oxford, and in London, he left academia to live by his pen and through such other work as he could obtain. This included a period as secretary of the London Zoo, as first director-general of UNESCO, and (first on radio and then on television) as a member of a very popular quiz show, *The Brains Trust*.

Huxley was always a keen evolutionist, and much of his writing was on and around the subject. His magnum opus, *Evolution: The Modern Synthesis*, was published in 1942, but it was the culmination of many years of similar writings and was followed by many more years of continued work. As the title suggests and as his career pattern confirms, Huxley was not so much a frontline empirical worker as a synthesizer. But given his wide interests, his many contacts, and his great skills as a professional writer (Huxley had collaborated with H. G. Wells and son on a massive popular treatise on biology), this was precisely the role for which he was suited. In this great synthetic work, he ardently intended to provide an essential foundation on which to build a first-class professional evolutionary biology.

Evolution: The Modern Synthesis (14) begins with a discussion of natural selection, following closely the argument given in the *Origin:*

> Darwin based his theory of natural selection on three observable facts of nature and two deductions from them. The first fact is the tendency of all organisms to increase in a geometrical ratio. The tendency of all organisms is due to the fact that offspring, in the early stages of their existence, are always more numerous than their parents; this holds good whether reproduction is sexual or asexual, by fission or by budding, by means of seeds, spores, or eggs. The second fact is that, in spite of this tendency to progressive increase, the numbers of a given species actually remain more or less constant.

From these two facts, we can deduce the struggle for existence. Then, we add in the third fact: the existence of widespread heritable variation. And with this and the struggle, we make our second deduction, to natural selection: "a higher proportion of individuals with favourable variations will on the average survive, a higher proportion of those with unfavourable variations will die or fail to reproduce themselves." Overall, and given time, evolution will ensue. Moreover, on balance, this will be evolution in the direction of adaptive improvement.

Next, Huxley turns to genetics, showing how natural selection and Mendelism are complements rather than rivals: how Mendelian genes can be passed on entire from generation to generation, and so selection can take effect and not be swamped out by blending of the units of heredity (or their effects); how mutations—that is, random variations—occur regularly but can be sufficiently small as not to upset the effects of selection but sufficiently big as to have an effect over the generations, as they mount up; how selection can act to promote diversity in populations as well as uniformity, thus collecting up ever-more variation; and much more.

Always Huxley, the oft-times popular writer, eschews mathematics and undue technicalities. He makes a point and then backs it with a series of judiciously chosen supports; no less than ten supports, for instance, underpin the existence and importance of small mutations. He begins with the fact that one can produce them artificially with x-rays, he works through such things as the polymorphism in natural populations, and he ends with the ways in which differences in populations can be correlated with the different conditions (and hence selective forces) encountered by subpopulations.

After presenting the central mechanisms of evolution, Huxley next turns (very much in the spirit of Darwin) to the application of his ideas. Of particular interest is the whole question of speciation: the breaking of interbreeding groups into daughter groups, isolated from their former fellow species members. Much time is spent on the issue of whether one always needs geographical isolation of populations, one from another, before one can get the formation of new species. Unlike the German-American ornithologist Ernst Mayr (1942), who at the time was writing that geographical isolation is critical to speciation, Huxley was inclined to think that on occasion ecological isolation (where organisms overlap geographically but occupy different though adjacent habitats) will suffice.

What about the other major area of evolutionary concern, paleontology? Unlike Darwin, Huxley had no firsthand experience of this subject; but he had always kept himself well-informed, beginning when he was young with discussion and information from Henry Fairfield Osborn, the leader among American paleontologists and an old student of his grandfather. Of particular interest was the question of trends in the fossil record, as when organisms get bigger over the years or develop more baroque forms. As a rule, Huxley saw little or no problem in fitting them into a Darwinian picture (494):

> The trends . . . would appear to present no difficulties to the selectionist, and it is hard to understand why they have been adduced as proof of nonadaptive and internally-determined ortho-

genesis. Whenever they are truly functional and lead to improvement in the mechanical or neural basis for some particular mode of life, they will confer advantage on their possessors and will come under the influence of selection; and a moment's reflection will show that such selection will continue to push the stock further and further along the line of development until a limit has been reached.

One point stressed by Huxley was that directed changes in time which do confer ever-greater adaptive advantage often come less through strife against the brute elements and more through competition between evolving lines (495):

> The evolution of the ungulates is not adapted merely to greater efficiency in securing and digesting grass and leaves. It did not take place in a biological vacuum, but in a world inhabited, *inter alia*, by carnivores. Accordingly, a large part of ungulate adaptation is relative to the fact of carnivorous enemies. This applies to their speed, and, in the case of the ruminants, to the elaborate arrangements for chewing the cud, permitting the food to be bolted in haste and chewed at leisure in safety. The relation between predator and prey in evolution is somewhat like that between methods of attack and defence in the evolution of war.

This is an important anticipation of what we shall see today's evolutionists calling "arms races," a metaphor drawing on the similarity between competition among animals and competition among nations that oppose one another.

We have now more than a flavor of Huxley's thought. Therefore let us turn to the kind of analysis that has been given of earlier thinkers.

Values of the Evolutionists

For all that I would qualify their contribution, I do not want to leave the impression that the population geneticists were indifferent to the epistemic necessities of good-quality science. That would be very far from the truth. Their sins, if such they be, were those of commission (adding the nonepistemic) rather than those of omission (neglecting the epistemic). Fisher had had good training both as a mathematician and as a theoretical physicist, and from this base he built a deserved reputation as the leading statistical theorist of his age. He used evidence to build predictions, he was concerned about consistency and consilience, his work was fertile in the extreme, and (as one would expect from a mathematician) he was extremely sensitive to simplicity and elegance.

Whatever his major motivation, when Fisher turned to evolutionary biology he was not about to forget epistemic values. Even though his work was essentially theoretical rather than experimental, epistemic factors were crucial. Most obviously, of course, was the question of consistency. All of this generation of theoreticians were eager to show that Darwinian selection and Mendelian genetics, far from being contradictory rivals, are complements. The one area of inquiry does what the other leaves open and demands. At the beginning of his book, *The Genetical Theory of Natural Selection,* Fisher was quite explicit on this score, showing how selection is the only mechanism that can dovetail smoothly with the Mendelian theory. A mechanism where the units of inheritance "blend" in each generation simply will not work. But more than this, even as he showed consistency, he was pushing the way toward other epistemic virtues—for instance, fertility—as he suggested that, for all that the synthesis creates or shows new gaps in our knowledge, it points to fruitful ways to plug them (20–21):

> The whole group of theories which ascribe to hypothetical physiological mechanisms, controlling the occurrence of mutations, a power of directing the course of evolution, must be set aside, once the blending theory of inheritance is abandoned. The sole surviving theory is that of Natural Selection, and it would appear impossible to avoid the conclusion that if any evolutionary phenomenon appears to be inexplicable on this theory, it must be accepted at present merely as one of the facts which in the present state of knowledge seems inexplicable. The investigator who faces this fact, as an unavoidable inference from what is now known of the nature of inheritance, will direct his inquiries confidently towards a study of the selective agencies at work throughout the life history of the group in their native habitats, rather than to speculations on the possible causes which influence their mutations.

Then, later in the book, Fisher demonstrated again and again that his theory has epistemic virtues, for instance, the ability to make predictions about what one should expect under certain specified circumstances. One example is his discussion of sex ratios, where he shows that these should balance themselves out so that parents are putting the same amount of effort into raising sons and daughters. If they do not, then the imbalance would lead to one group getting an advantage over others, which would continue until they competed against each other, thus putting matters to rights: "Selection would thus raise the sex-ratio until the expenditure upon males became equal to that upon females" (143). This speculation about sex ratios has been one of the most fruitful ideas in the history of evolutionary theory, with huge amounts of predictively fertile (and confirmed) work following from it. Whatever the cultural elements in Fisher's work— which generally reduce today's evolutionists to embarrassed silence—its epistemic virtues have more than stood the test of time.

Interestingly, when we come to Huxley's *Evolution: The Modern Synthesis*, published some twelve years later, the epistemic factors were, if anything, in decline. I do not want to exaggerate. Huxley was eager to conserve the advances made by Fisher and the other population geneticists. For instance, he too makes much of the consistency between natural selection and Mendelian theory, linking this indeed to a second virtue, simplicity. "The rise of Mendelism, far from being antagonistic to Darwinian views (as was claimed, notably by the early Mendelians themselves, in the years immediately following its rediscovery), makes a selectionist interpretation of evolution far simpler" (55). Moreover, in some respects Huxley struck out on his own, showing the fertility of his ideas in directions not covered by Fisher. The work on arms races is a case in point. And whereas Fisher admitted candidly that he did not deal with (nor was he desperately interested in) evolution beyond the causal mechanisms and their immediate effects today, Huxley aimed to give a broad sweep of the whole evolutionary picture, as Darwin had done. After all, that is what a synthesis is all about: bringing many different areas together under one causal process.

But when all is said and done, perhaps precisely because Huxley does give a synthesis—a survey—rather than a report of frontline research, one rarely gets an overwhelming conviction of epistemic excellence from his work. Arms races notwithstanding, one does not usually get the sense that one can pick up from Huxley's work and make predictions and fertile advances into new territory. At the least, one needs to turn to the work on which he is reporting. Likewise with simplicity. There are some elegantly simple ideas, but they are Darwin's, not Huxley's. For all that he lauded simplicity, his own style does not produce compelling, beautiful hypotheses. Rather, Huxley piles up one piece of evidence after another until the reader collapses with exhaustion or (dare one say) boredom.

Perhaps Huxley's forays into science journalism with the Wellses backfired, setting up bad habits or a willingness to settle for less than fully mature science. But whatever the reason, judging now from our perspective, one does get the strong sense that Huxley was not offering a work truly intended to be a tool of research— with all the essential marks of a Kuhnian paradigm, as one might say. Which at once raises the question of whether there were other things driving Julian Huxley—things in the nonepistemic realm, perhaps.

We can indeed easily discern a range of values drawn from the culture of Huxley's own day that found its way into his theorizing. The arms race picture, for instance, drew on the military technology of the first part of the century. Given the full and detailed way in which Huxley described (and went on describing) the new and rapidly changing technology of modern warfare, one does indeed have the feeling that he has moved from disinterested description to enthusiastic approval. Barbed wire and machine guns come in for praise: "Advance is so great that an entire method of attack or defence is rendered obsolete" (495).

But this and everything else paled beside the main force motivating Huxley: our old friend, progress. With the possible exception of Herbert Spencer, no one in the whole history of evolutionism was more ardent in his progressionism than Julian Huxley. He lived it, breathed it, talked it, and wrote about it at very great length. Searching desperately as a young man for a faith to substitute for Christianity, Huxley found it in progress—and for him, progress was best manifested in and made most probable and plausible by the evolutionary process. Particularly important here was an early reading of *Creative Evolution* by the French philosopher Henri Bergson. Believing that a life force, the *élan vital*, motivates and drives organisms, Bergson argued that there is an upward cast to the history of life. Realizing that such vitalism is never in itself going to be an adequate basis for genuine science, Huxley downplayed the metaphysics: "Bergson's *élan vital* can serve as a symbolic description of the thrust of life during its evolution, but not as a scientific explanation" (457–458). Yet, although for metaphysical vitalism he substituted Darwinism (after a fashion), Huxley hoped always to keep vitalism's central message of upward progress. For Huxley, Bergson's inadequacies as a scientist were a challenge, the beginning of inquiry, not the end of the debate. Indeed, *Evolution: The Modern Synthesis* itself treats the rest of the evolutionary discussion as a prolegomenon to the concluding sections where finally Huxley gets to progress, making very clear that the intellectual climax mirrors the biological climax, that which was marked by the appearance of our own species, *Homo sapiens.*

Human progress, however, involves the control of the environment and is marked by extreme cultural complexity and a division of labor; it is essentially different from the limited progress—a kind of specialization—that we find lower down the evolutionary scale. In a sort of reversal of form, the key to the success of humans is that we have remained generalists and so have a flexibility that is denied other organisms. Moreover, now that we have succeeded, we have cut off the chances of any others to rise up and challenge us. Our progress, which opens the door to unlimited opportunities in the cultural realm, comes at the expense of the opportunities of other species. It is indeed our unique possession (570):

> The last step yet taken in evolutionary progress, and the only one to hold out the promise of unlimited (or indeed of any further) progress in the evolutionary future, is the degree of intelligence which involves true speech and conceptual thought: and it is found exclusively in man. This, however, could only arise in a monotocous mammal of terrestrial habit, but arboreal for most of its mammalian ancestry. All other known groups of animals, except the ancestral line of this kind of mammal, are ruled out. Conceptual thought is not merely found exclusively in man: it could not have been evolved on earth except in man.

By this stage of his work, Huxley has abandoned most pretenses of being particularly Darwinian. Indeed, as in those whom he criticized, in Huxley evolution had now gained its own internal momentum. In fact, one might fairly say that the biggest mistake in understanding Huxley would be to think that he ever truly turned his back on vitalism. He showed little interest in producing full-blooded progress as the end result of an arms race. The impression one gains from Huxley is that life would keep going up no matter what the biological state of affairs might be. At the final point, progress had transcended its material basis.

Huxley's Science in His Own Time

Obviously in speaking of Huxley's being influenced by cultural factors, I am not implying that he was drawing on ideas which were necessarily shared and appreciated by everyone in his society. Indeed, some people found his progressionism so offensive that when he wrote a manifesto for UNESCO incorporating his philosophy (Huxley 1948), critics took it as an opportunity to deny him a full four-year term as director. But Huxley's progressionism was clearly part of his cultural milieu, and its influence on his scientific theorizing cannot be denied, nor should it be minimized.

How did people of his day regard the biology of *Evolution: The Modern Synthesis*? One thing we can say with some certainty is that it was Huxley's hope and intention that it be regarded as the secure platform on which a new, mature, fully professional evolutionism could be constructed. A sort of *Origin of Species* of the twentieth century, as much ahead of Darwin as this century is ahead of the last. To quote the final words of the first chapter: "It is with this reborn Darwinism, this mutated phoenix risen from the ashes of the pyre kindled by men so unlike as Bateson and Bergson, that I propose to deal in succeeding chapters of this book" (28).

This was a confident position to take—arrogant perhaps—but it was not entirely out of place. Huxley began life with great advantages. The family connections were massive. He was the grandson of the great T. H. Huxley as well as the great-grandson of Thomas Arnold of Rugby, the dominant figure in nineteenth-century secondary education and father of Matthew Arnold, the poet and essayist. With this was combined the boost of training at the pinnacles of England's school and university education. Through his life, these sorts of factors paid off. Huxley was born at the center of the scientific network. He never had to struggle to gain attention. He was elected to the Royal Society almost as a matter of right. He was acknowledged as one of Britain's leading intellectuals; in 1930 *The Spectator* had rated Huxley one of Britain's five best brains (Kevles 1992, 241). Nor did it hurt that Julian's brother, Aldous, author of *Brave New World,* was one of the truly notorious and most widely read novelists of the day.

Add to this the fact that, even if he held no university post for most of his life, Julian Huxley worked unceasingly—at his writings, at public affairs, and very much at organizing things in the scientific community. He got a first at Oxford, not to mention winning a prize for poetry. He did do some interesting work on animal behavior and (later) on problems of development (Huxley 1932, 1934). Most importantly, it was he who, around 1940, was urging evolutionists in both Britain and America to start organizing and doing those things that one finds in successful scientific disciplines: mentoring students, seeking grants, and founding and maintaining journals. As it happens, this last project came to naught in the Old World, but it was a strong stimulus on those in North America who had similar aspirations.

Into Professional Oblivion

Yet, even before we turn to the intellectual factors, we should note that clouds were on the horizon. With its advantages, his heritage brought its disadvantages also. Like his grandfather, Huxley was subject to crushing depressions ("nervous breakdowns"), which in his case prevented him from ever sticking systematically to tasks, like building a research group that could go forth and develop the paradigm he was promoting. To this was added too much devotion to self and not enough to the sensibilities of those around. He was fired from his post as secretary of the zoo essentially because he was following his own interests rather than those for which he was paid. Added to that were the UNESCO troubles.

But most fatally, Huxley, although clever, was simply not of the first rank as a scientist and never built a record of sustained creative achievement. He got his F.R.S. mainly through connections; and in the eyes of his contemporaries, *Evolution: The Modern Synthesis* always had the fatal air of being no more than it claimed: a synthesis and not something at the cutting edge. It was a useful book, but it was not very exciting. Notwithstanding bits and pieces (such as the arms race hypothesis), generally Huxley's work predicted nothing special in its own right, nor bridged new gaps showing consistency and coherence (this had been done ten or more years before by the population geneticists), nor pointed the way forward into fertile fields of new inquiry—the sort of thing a bright graduate student could take up. It was more useful in the preparation for comprehensive examinations than in the search for a thesis topic (Baker 1976).

And then there was the question of Huxley's cultural commitments, most particularly that of progress. We are not now talking about metavalues but about cultural values built right into the science. Even in Huxley's grandfather's day there had been something wrong about suggesting that top-quality science might confirm or prove the essential rightness of so obvious a cultural value as progress. Things were no different as the twentieth century moved to midpoint. A price had to be paid: if one insisted on retaining progress, then either one had to find an epistemically secure notion of progress or one had to forgo hopes of top-quality science.

Unfortunately, Julian Huxley hoped to have his cake and eat it too. *Evolution: The Modern Synthesis* had started as an address to the British Association for the Advancement of Science (Huxley 1936). That was acceptable, for the BAAS was a forum for the general public where one was expected to comment on the broader aspects of science—especially the broader cultural aspects of science. But when Huxley expanded his talk to make a would-be foundation for an ongoing research program, professional scientists wanted nothing of it. Despite his distinguished position, he was kept out of the journals and refused grants. Even the philosophers joined in the nay-saying when Huxley tried, in Spencer-fashion, to use evolution as the basis of ethics (Broad 1949).

The climax came in the late 1950s, by which time Huxley had gone so far as to endorse the world picture of the French Jesuit paleontologist Pierre Teilhard de Chardin (1955), who had argued that the evolution of life progresses up to the "omega point," something which the priest identified with Jesus Christ. The Nobel Prize–winning biologist Peter Medawar (1961) wrote an absolutely savage critique of Teilhard, altogether dismissing him as deserving serious attention: *"The Phenomenon of Man* [Teilhard's magnum opus] cannot be read without a feeling of suffocation, a gasping and flailing around for sense" (71). To Medawar (as to others) it was simply incomprehensible that some people, notably Huxley, who wrote the preface to the English translation, would let themselves be deceived by such nonsense. "If it were an innocent, passive gullibility it would be excusable; but all too clearly, alas, it is an active willingness to be deceived" (81).

Nothing more could be said after this. People were happy to give Huxley honorary degrees on the occasion of the centenary of the *Origin of Species*. They, evolutionists especially, wanted nothing to do with him when it came to real, professional science. Whether it be true that the best science is truly culture-free, by the 1940s and 1950s it was a universal demand of the internal culture of science that science be external-culture-free. Because he broke the code, flagrantly, Julian Huxley failed to make the desired breakthrough from popular to fully mature professional science. And that was the judgment on his evolutionism.

5

><

THEODOSIUS DOBZHANSKY

Evolution Comes of Age

Evolutionists kept their distance from Huxley for good reason. Science in the twentieth century is never done in isolation. Others are always circling, wanting to grab the goodies: students, grants, research space, places in the curriculum, and more. Which evolutionist has not faced demands that his or her course be dropped because students need one more course in biochemistry, better to qualify for medical school? In the 1920s and 1930s, experimental embryologists were the big threat. Later, the molecular biologists took the role as main rivals. In 1965, by which time some evolutionists had made slight inroads into Oxford University, the only professorship they had managed to get their hands on was taken away from them and given to a molecular biologist—much at the urging of Peter Medawar! People would rightly have dreaded a public visit from Huxley.

Yet by the 1960s, evolution was rising up the scale of respectability, socially and epistemologically. This was happening in Britain and even more so in America. So, let us cross the Atlantic and see how another person managed to succeed where Huxley failed.

Genetics and the Origin of Species

Theodosius Gregorievitch Dobzhansky was born in Russia in 1900 and died in 1975, the same year as Julian Huxley. Determined from a young age to be a biologist, he survived the Revolution and trained as a field naturalist (studying ladybirds) and as a geneticist. Sent to America in the late 1920s to study with T. H. Morgan, he settled and lived there for the rest of his life. Switching to the hot organism of the day, the so-called fruit fly *Drosophila*, Dobzhansky did skilled chromosomal studies and then (increasingly) field work, looking at populations of wild flies and at how they varied in space and time. In 1936 he gave a series of prestigious lectures in New York, and then these were written up rapidly and published a year later as *Genetics and the Origin of Species* (1937).

Much of the appeal and strength of Dobzhansky's book came thanks to his Russian background, for he was able to draw on that country's naturalist tradition in a way virtually unknown to native-born Americans. He knew from firsthand experience that masses of variation existed in the wild; he knew that one gets gradual transitions from one form to another; he knew that adaptation is a real phenomenon and that (given the untenability of Lamarckism) Darwin's ideas alone make a concerted effort to explain it. "A biologist has no right to close his eyes to the fact that the precarious balance between a living being and its environment must be preserved by some mechanism or mechanisms if life is to endure. No coherent attempts to account for the origin of adaptations other than the theory of natural selection and the theory of the inheritance of acquired characteristics have ever been proposed" (150).

The heart of Dobzhansky's book, however, the framework which made sense of everything, was American; it was Sewall Wright's shifting balance theory, which Dobzhansky drew on and used as the crucial central mechanism of evolutionary change. Most particularly, it was the metaphor of an adaptive landscape with its peaks and valleys, with groups of organisms occupying the high ground or undergoing processes which lead to movement from one peak to another (187):

> Each living species or race may be thought of as occupying one of the available peaks in the field of gene combinations. The evolutionary possibilities are twofold. First, a change in the environment may make the old genotypes less fit than they were before. Symbolically we may say that the "field" has changed, some of the old peaks have been leveled off, and some of the old valleys or pits

have risen to become peaks. The species may either become extinct, or it may reconstruct its genotype to arrive at the gene
combinations that represent the new "peaks." The second type of
evolution is for a species to find its way from one of the adaptive
peaks to the others in the available field, which may be conceived
as remaining relatively constant in its general relief.

Dobzhansky wrote of species as "exploring" the land around a peak in a
kind of "trial and error" process, until it found its way to move across a
valley and up the slope of another peak.

    One should note here an important difference between the atti-
tudes of Julian Huxley and Dobzhansky with respect to the formalisms
of the population geneticists. Neither Huxley nor Dobzhansky had any
mathematical ability whatsoever—the technical work of Fisher and of
Wright was a closed book to both of them. However, whereas Huxley
simply ignored the work of Fisher and drew on his own sources,
Dobzhansky (thanks to the simple metaphor of an adaptive landscape)
was able to grasp and use Wright's basic theory. The mathematics was
taken out, but the main ideas were presented and then used by
Dobzhansky to explain the facts of nature as he saw them.

    Perhaps in part precisely because he was a working biologist en-
thusiastically using the mechanism to which he was committed,
Dobzhansky was not into the synthesizing business as was Huxley.
Dobzhansky dealt with real change and variation in real populations as
they occur today, and left matters at that. Nothing on paleontology, for
instance. So in *Genetics and the Origin of Species*, following on the
introduction of Wright's model, we get detailed discussion of the ways
in which species might be formed—including special cases where chromosomal factors become paramount and not-so-special cases where
species simply break apart and selection perfects the mechanisms that
isolate species, reproductively, from one another. Was not this all rather
limiting, with the focus on the present, precluding important discussions about the events of the past? Apparently not, for it was ever
Dobzhansky's creed that the very large and long can be explained by
the very small and short: "Experience seems to show . . . that there is
no way toward an understanding of the mechanisms of macro-evolu-

tionary changes, which require time on a geological scale, other than through a full comprehension of the micro-evolutionary processes observable within the span of a human lifetime and often controlled by man's will" (12).

Genetics and the Origin of Species proved to be a very important work. Person after person read it and found within it the reasons for moving into and forward with evolutionary studies. Sociologically, it functioned very much in the way that Kuhn supposed for a work which founds a paradigm. Hence, Dobzhansky himself moved at once to the front of evolutionism in America, a position he held until his death. Not that one should conclude that Dobzhansky remained idle or static in his thinking. Genetic drift was a key component in Wright's formulation of his balance theory, and as such was accepted by Dobzhansky in the first edition of *GOS*. But in the early years after this edition was published, field work (then backed by experiments) convinced Dobzhansky that drift is far less significant than he had thought (Dobzhansky 1943; Wright and Dobzhansky 1946; see figure).

The reason lay in the changing nature and structure, within *Drosophila*, of the chromosomes, those gene-carrying stringlike entities in the nucleus of cells. That there might be differences in chromosome patterns between members of populations was one thing: indeed, given that changes in chromosome pattern can have an effect similar to changes in gene pattern (known as mutations), and given also his belief in variation, Dobzhansky virtually expected such differences. What he did not expect is that, within the same population, one would find systematic cyclical changes of percentages of specific chromosome patterns from one form to another at one season of the year and then back to the first form at another season of the year. Such regularity of change belied the possibility that a random nondirected mechanism like drift was responsible for variation. One needed a mechanism which can sustain a pattern, and this he found in natural selection.

One cannot stop here. If what is happening with the fruit flies has a more general message for all organisms, namely, that one should think primarily in terms of natural selection rather than drift—and this was a message that Dobzhansky read from his observations and experiments—then this triggers a search for mechanisms that might, as a

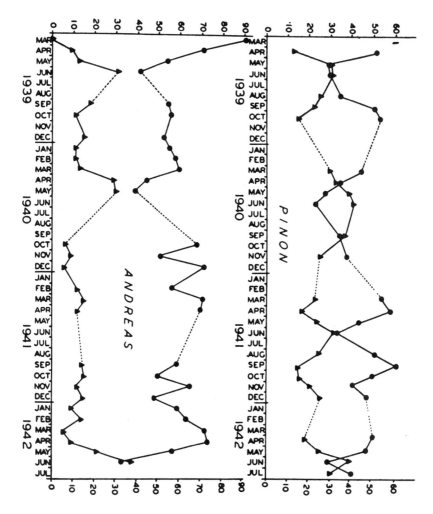

The cyclical patterns of variation in chromosome arrangements, the same in different populations, which convinced Dobzhansky that selection matters more than drift (from Dobzhansky 1943).

matter of course, produce and preserve significant genetic variation within natural populations. Variation from mutation alone could probably provide enough raw material for evolution, were it not for the fact that selection eliminates the very variation on which it depends. By moving a species toward adaptation, it seems to put itself out of business; it is self-destructive. Of course, systematic changes in physical circumstance, such as the climatic changes that seemed to be responsible for changes in the chromosome structures of *Drosophila*, would be one way in which selection could fail to eliminate all variation. As the circumstances change endlessly, so does selection reverse itself endlessly. But one cannot expect that this is a regular occurrence.

Something more is needed to supply unending variation, and eventually the answer Dobzhansky decided upon was "balanced superior heterozygote fitness" (Dobzhansky and Wallace 1953; Dobzhansky and Levene 1955). Thanks particularly to this mechanism, there are always masses of variation within a population. To use a metaphor: the conventional view (that variation is the result of the occasional mutation, which will probably be no use anyway) is analogous to needing a book for an essay and having to wait each month for the main offering of the Book-of-the-Month Club—almost certainly a selection which will not meet your needs. Dobzhansky's view is that the process of natural selection ensures that a population keeps a whole library always at your disposal. You have to write an essay on dictators? Well, if there is nothing on Napoleon, there may well be something on Hitler. And if not Hitler, then Stalin. Selection is not guaranteed one unique perfect offering, but there is usually a range of possibilities.

And this is the way things are in real life. Faced with a new predator, a species may have no unique perfect solution. Of the organisms that survive to reproduce, some may owe their success to more efficient camouflage, others may have behavioral tendencies that helped them elude the predator, and yet others may have developed more effective armor (Ruse 1982). Over time, these differences will be magnified into separate species. This kind of thinking had great effect (see box).

We are getting to the point in our story where past and present meld into one; but for the moment let us stay with our practice of looking first at matters ahistorically, and then relating the discussion back to the events of the day.

Values in Dobzhansky's Science

The move from Huxley to Dobzhansky was from night to day. Many of their values overlapped, but how these values functioned in their

Balance

Natural selection changes the genetic composition (the *gene pool*) of populations of organisms, but sometimes it can act to maintain genetic variation. One of the most discussed ways is through superior heterozygote fitness *(heterosis)*. Suppose at some locus within a (large) population one has two alleles, *A* and *a*. This gives rise to three possibilities, the two homozygotes (*AA* and *aa*) and the heterozygote *(Aa)*. Suppose now that the heterozygote is fitter than either homozygote. This means that the *Aa* genotype is always going to produce more offspring than either the *AA* genotype or the *aa* genotype, and so even if either or both of the homozygotes produce no offspring at all, thanks to the Hardy-Weinberg law both alleles and all three genotypes will be represented in the next generation.

Strictly speaking, this does not mean that there will necessarily be a balance—an equilibrium of allele and genotype ratios—but in fact one can readily show that such a balance will obtain. Selection is acting to preserve the status quo, rather than destroy it.

The best-documented case of balanced superior heterozygote fitness is found in humans, and it centers on sickle cell anemia. In parts of Africa about 5 percent of each generation die in childhood from this inherited form of anemia, which is now known to be a function of homozygosity for a certain mutant allele. However, these deaths are balanced in the population because heterozygotes for the gene have an innate protection against the effects of another killer, malaria. The heterozygote is fitter than the homozygote without the sickle cell gene at all. Nature purchases the superior health of some at the cost of the deaths of others.

science is another matter. Start with the question of prediction, or rather of prediction-cum-fertility, and consider the work that Dobzhansky himself built on *Genetics and the Origin of Species* in the years after the first edition of the book. The discovery of rapid changes in the chromosome pattern of populations of fruit flies, and the cyclical regularity of these changes from one form to another and then back again, altogether precluded an explanation based on drift or any other accidental force: "The available data seem to fit best a . . . hypothesis, which assumes that the carriers of different gene arrangements in the third chromosome have different ecological optima. At Andreas Canyon and at Pinon Flats the Standard gene arrangement is favored in spring and the Chiricahua arrangement in early summer. Natural selection alters the composition of the populations accordingly" (Dobzhansky 1943, 175–177).

At once, Dobzhansky was led to make predictions about what one would expect if one tried to replicate nature in the laboratory (Wright and Dobzhansky 1946, 154–155):

> The experimental results demonstrate clearly that there may be selective differences between chromosome types derived from the same locality and that these may be of such a nature as to result in the indefinite persistence of several types in such a locality. The marked rise in frequency of Standard and decrease of Chiricahua in summer, observed to occur in nature in the locality from which the flies were collected, is analogous to the experimental reaction to high temperature, and it is tempting to compare the lack of change in artificial population at 16.5 with the constancy observed in the natural ones during fall and winter.

More than just this: Dobzhansky set up a whole research tradition whereby people could strive for predictive understanding at the causal level. Consider the variation claim: that when new needs arise, appropriate variations are already available among the individuals in a given population, obviating the need to wait on rare and random mutations. Dobzhansky's student Francisco Ayala used this theory to predict that populations of fruit flies, although generally alcohol-intolerant, should carry enough variation that they could evolve in the direction of tolerance, should the need or opportunity arise. Experiment triumphantly confirmed this prediction. Within but a very few generations of artificial selection, strong alcohol tolerance was developed (Ayala et al., 1974). In the wild, *Drosophila* in the vicinity of wineries positively thrive on what would kill their abstemious conspecifics.

This is prediction. Other epistemic values were likewise satisfied, not least when fellow evolutionists, like the ornithologist-systematist

Ernst Mayr (1942) and the paleontologist George Gaylord Simpson (1944), extended Dobzhansky's theory into other areas. It was possible to achieve a sound scientific understanding of all sorts of phenomena in biogeography, paleontology, and other branches of biology. I certainly do not want to say that evolutionary biology achieved great heights overnight. As the evolutionists themselves openly admitted, their work was hardly as predictively hard-nosed as that of the physicists. Even where success was achieved, qualifications and admissions of ignorance were usually required. "It appears that there are important factors yet to be discovered" (Wright and Dobzhansky 1946, 155). The point I want to make is that the epistemic standing of evolutionary theory after the 1930s was as much above the standing of such work after the *Origin* as that book had been above the work to be found in Erasmus Darwin and his contemporaries.

Now, what do we say on the cultural front? Theodosius Dobzhansky was a deeply emotional man, torn by a love for Russia as he had known it and a horror at the history through which it was now passing; immensely grateful to America for the opportunities that it offered him and his family but desperately worried also about the threat of war and (in the 1950s) the possibility that America might launch a nuclear attack on his birthland; overwhelmingly attached to his students and his friends but ready to take umbrage at the smallest possible slip or perceived slight; and much much more. It is simply inconceivable that such a man could have been unmoved by his culture, or rather his cultures. Nor was he. From the first, Dobzhansky admitted that he went into the evolutionary quest with a mission—in his case religious, with a hope to show that God is working His purpose out in the evolutionary realm and that man is His finest creation, the apotheosis of an upwardly reaching, progressive life process (Greene and Ruse 1996).

In the early years Dobzhansky found all that he needed within the Russian tradition, which, being much influenced by German transcendentalism, was teleologically human-directed through and through. When he moved to America, he turned more to Bergson and those who at that time were pushing related philosophies, men such as Whitehead. Later, like Huxley, Dobzhansky also discovered Teilhard and began endorsing his ideas publicly, with enthusiasm. Although raised in the

Russian Orthodox Church, Dobzhansky was happy to embrace a kind of pan-Christianity, with the whole of nature pointing upward to the godhead reincarnate.

Dobzhansky was quite open in his feeling that—however he would treat of progress in his formal science—such science would, in his eyes, lose many of its attractions if it could not underpin his progressionism. Bergson and Teilhard were no help here. But the adaptive landscape metaphor of Wright's shifting balance theory was another matter—hardly surprising when one considers how it played a similar progressionist role for Wright himself. By making some peaks higher than others and by implying that change in the landscape is going to be slow or virtually nonexistent, the landscape metaphor readily takes on a progressionist reading: a reading that was at the heart of Dobzhansky's vision of evolution.

Other nonepistemic values throve alongside and entwined with this progressionism. Take the question of variation and the hypothesis of balanced superior heterozygote fitness. Where did Dobzhansky get this idea, or, more particularly, where did he get the idea that it might be a significant factor in the real world? The balance hypothesis itself apparently came from his friend I. Michael Lerner (1954), who was in turn inspired by holistic ideas of the 1930s suggesting that new properties emerge when parts are put together (that is, wholes become greater than the sum of their parts). That such new properties do emerge is something that can be put to ready test. But why should one think them superior and widespread?

Within Dobzhansky's hypothesis, this was the key element—an element bitterly opposed by the Nobel Prize–winning geneticist Hermann J. Muller and his students. They denied strenuously that populations show the variation that Dobzhansky supposed and needed. To use the labels that were given to the two sides, they denied the "balance" position while endorsing the alternative "classical" position, that mutations are rare and either on their way to complete success or to extinction (Muller 1949; Muller and Falk 1961; see box).

The cultural underpinnings of these two sides, at dispute in the darkest days of the Cold War, were there for all to see. Muller, an ex-communist, was violently opposed to nuclear weaponry. As part of

Two Hypotheses

Balanced superior heterozygote fitness does undoubtedly exist in nature. But one swallow does not make a summer, and herein lies the debate between Dobzhansky and Muller. The former thought that such balance and like selection-promoting mechanisms were the norm, and the latter thought them rare. Hence the former saw lots of variation in populations, and the latter saw very little. Diagrammatically, one can show the rival hypotheses as follows:

	Individual 1	Individual 2
Balance	$A_1 \, B_3 \, C \, D_1 \, E_1 \ldots Z_1$	$A_3 \, B_5 \, C \, D_2 \, E_1 \ldots Z_5$
	$A_2 \, B_4 \, C \, D_2 \, E_2 \ldots Z_2$	$A_4 \, B_8 \, C \, D_2 \, E_1 \ldots Z_7$
Classical	$A_1 \, B_1 \, C \, D_1 \, E_1 \ldots Z_1$	$A_1 \, B_1 \, C \, D_2 \, E_1 \ldots Z_1$
	$A_2 \, B_1 \, C \, D_2 \, E_1 \ldots Z_1$	$A_1 \, B_1 \, C \, D_2 \, E_1 \ldots Z_1$

Part of what makes it so difficult to decide between the two hypotheses is the overlap between them. Balance supporters agree that there will be no variation at all at some loci (C in this example) and that selection will be moving new mutants up or down at some loci (E in this example). Conversely, not only do classical supporters see some cases where selection is moving mutants up or down (A in this example) but they accept balanced variation at (a few) other loci (D in this example).

his opposition, he argued that nuclear tests, putting radiation into the air, are deleterious for us all and most especially for the children of future generations. Dobzhansky, though he hated war, nevertheless thought that the West must maintain its nuclear superiority and that if this means testing, then so be it. He and his students were therefore keen to show that the radiation artificially introduced into the atmosphere has little or no bad effect—possibly a good effect even! The balanced heterozygote fitness mechanism was tailor-made for this end: there is always a lot of variation in the population, and a few more radiation-induced mutations will not make much difference.

I do not in any sense suggest that Dobzhansky and students fraudulently altered results to get the desired answers. But I have little doubt that cultural factors lying behind the formal science that Dobzhansky was producing helped significantly to flesh out the gaps between the proven and the presumed. The fact that the Atomic Energy Commission was delighted with these results and happy to support the work of Dobzhansky and his students was a nice bonus. Everybody's ends were being served (Dobzhansky and Wallace 1953, 1959).

Like Darwin before him, Dobzhansky felt able to expand out from his beliefs in progress to a whole range of other cultural values which he saw as being endorsed directly by his science. Dobzhansky's later writings contain much about freedom, morality, and religion. Of particular concern to him was the nonequivalence of biological identity and human worth or value. Dobzhansky kept coming back to this idea, for it is in our individuality that we find the essence of being human: "The genes conditioning the variations in the appearance, physique, intelligence, temperament, special abilities—in short the genes making people recognizably different persons, really unique and nonrecurrent individuals—these genes may be maintained in human populations in balance states, either because of being advantageous in heterozygotes or because of the action of diversifying selection" (Dobzhansky 1962, 298). Here, he stressed, we have simply those "differences of the sort we observe among healthy or 'normal' people," and there is no reason at all to think any one better in some absolute sense than any other.

The other evolutionists in the school Dobzhansky founded— Mayr, Simpson, and the botanist G. L. Stebbins—were not far behind the leader in their public enthusiasm for the nonepistemic, the cultural. Simpson, for instance, was forever writing about the virtues of democracy and how this is connected with evolution as he saw it. His views differed significantly from those of his close friend Julian Huxley (1943), who favored large-scale public works and other state-funded projects; Simpson (1949) looked for more from the individual. For the paleontologist, there were two major directives. First, there was the need to improve and promote knowledge—knowledge in itself, as a good (311):

> The most essential material factor in the new evolution seems to be just this: knowledge, together, necessarily, with its spread and inheritance. As a first proposition of evolutionary ethics derived from specifically human evolution, it is submitted that promotion of knowledge is essentially good. This is a basic material ethic.

"Promotion" involves both the acquisition of new truths or of closer approximations to truth (metaphorically the mutations of the new evolution) and also its spread by communication to others and by their acceptance and learning of it (metaphorically its heredity).

Then, second, we have personal accountability, which leads to integrity and dignity (315):

> The responsibility is basically personal and becomes social only as it is extended in society among the individuals composing the social unit. It is correlated with another human evolutionary characteristic, that of high individualization. From this relationship arises the ethical judgment that it is good, right, and moral to recognize the integrity and dignity of the individual and to promote the realization or fulfilment of individual capacities. It is bad, wrong, and immoral to fail in such recognition or to impede such fulfilment. This ethic applies first of all to the individual himself and to the integration and development of his own personality. It extends farther to his social group and to all mankind.

This valuing of responsibility and dignity was very much a function of the times and society within which Simpson lived. We are talking now not only of the years when the Cold War had settled into the long winter but specifically of the time when Soviet science was suffering under influential charlatans. Notable here was the agriculturalist Trofim Denisovitch Lysenko, who had the ear of Stalin because he claimed to be able to speed up through non-Mendelian means the production of more and better quality wheat and who used his power to oppress or murder those biological opponents who dared to point out how fraudulent this all was (Joravsky 1970). For biologists like Simpson, this persecution of genuine scientists made the crusade for freedom and democracy a personal issue.

It is therefore not surprising that from these evolutionary speculations about dignity and responsibility, Simpson launched straight into a condemnation of the oppressive regimes then flourishing in the East, and he juxtaposed this with a cherishing of the society within which he found himself: "Democracy is wrong in many of its current aspects and under some current definitions, but democracy is the only political ideology which can be made to embrace an ethically good society by the standards of ethics here maintained" (321). This was then all tied in with a biological basis for ethical action: "It bears repeating that the evolutionary functioning of ethics depends on man's capacity, unique at least in degree, of predicting the results of his actions. A system of naturalistic ethics then demands acceptance of individual responsibility for those results, and this in fact is the basis for the origin and function of the moral sense" (145–146).

Material similar to that of Simpson can be found in the popular writings of Mayr (1988) and Stebbins (1969). The latter, during the late 1960s, while a faculty member at Berkeley, even went so far as to argue that it is good biologically to have a small group of radicals upsetting an otherwise complacent population! The links between biology and morality are not always as tight and explicit as one might desire (given the promissory notes that everyone offered), but ultimately, supposedly, it is possible to relate all the moral and like sentiments back to elements within the best science of the new brand of evolutionary theorizers.

Exactly how these elements (presumably cultural) related to the science needs a little more unpacking, but at this point it is most useful to move the clock back fifty years and ask about the status and nature of the work in evolution at the time Dobzhansky and his friends and colleagues were active.

Dobzhansky's Science in His Own Time

One has to say that on the social level the American evolutionists did work hard—and a great deal more successfully than Huxley—to upgrade their science from the popular level of the previous three quarters of a century. They wanted a fully mature, professional science in which they and their students could spend a lifetime of research; they knew the sorts of things that such a science requires; and they schemed and pushed and labored to secure them (Ruse 1996).

When he wrote his big book, Dobzhansky (unlike Huxley) was a practicing biologist. Located in Morgan's laboratory, he was at the heart of the science as it was in the mid-1930s. This got him respect and status

in a way that was not there for Huxley—and his reputation was sol-
idified in 1940 when Dobzhansky became a professor at Columbia
University. Add to this that Dobzhansky was a genius at infecting
others with his enthusiasm and that he knew how to compensate for his
own lack of mathematical training. Very much at his instigation, the
seminal papers (after the first edition of *Genetics and the Origin of Species*)
on the effects of selection were written in collaboration with Sewall
Wright—Dobzhansky doing the empirical work and Wright doing the
calculations—so that science of a formal, mathematical kind, the high-
est-seeming kind, got done there also. This picked up later as Dobzhansky (again, very much unlike Huxley) gathered to himself a steady
stream of brilliant students, whom he cherished and treated like children (his own term) and who would work with him carrying forward his
program in mathematically sophisticated ways quite beyond the ken of
their leader.

Parallel with this, Dobzhansky found people willing to fill in the
gaps of the evolutionary picture as he had sketched it. Here, through
good fortune and perseverance, he succeeded mightily well. The col-
laborators Mayr, Simpson, and Stebbins did not appear by chance. First,
fellow immigrant Ernst Mayr was coaxed into writing a work on systematics and ornithology from the Dobzhansky perspective. Then, two
years after Mayr's *Systematics and the Origin of Species* (1942), the brilliant
mammalian paleontologist George Gaylord Simpson gave a fossil record reading of the picture: *Tempo and Mode in Evolution* (1944). Finally,
at the end of the decade (1950), his first choice having let him down,
Dobzhansky persuaded G. L. Stebbins to write the botanical arm of the
synthetic theory: *Variation and Evolution in Plants.* This time Dobzhansky had taken no chances, inviting Stebbins to live with him while he
was working on the book and discussing precisely the ideas which
should be included!

There were other things. A society was founded, a journal (*Evolution*) was started, and then, at the beginning of the 1950s, the National
Science Foundation began to provide grants. Dobzhansky and his
friends were at the head of the line—putting in proposals, writing
strong reviews of one another's applications, being as negative as possible about the opposition. They schemed to get co-workers and sympa-

thizers and then students elected as members of the National Academy of Sciences. In short, they did all of the time-honored things that lead to professional success.

But what of the epistemic and nonepistemic factors in the work of Dobzhansky and friends? On the epistemic front, it was precisely the kinds of values we have highlighted that persuaded people—within and without evolution—that things had changed from days of yore. When it came to the norms of science, Dobzhansky showed a far deeper and more subtle sense of their significance than did Huxley—and his fellow biologists appreciated this fact. Dobzhansky provided, in the best Kuhnian style, something with which would-be full-time evolutionists could work. No one was about to complain about the lack of mathematics since, with the exception of Simpson, who was truly gifted at those sorts of things, no one else could follow the mathematics either. But they could follow Wright's landscape metaphor and they could see how Dobzhansky had taken it and used it as a tool of research. Others too, students and colleagues and co-workers in the field, could follow suit.

After Dobzhansky, evolutionists really had something to do and a way to do it. More than this, Dobzhansky led the way himself. Not only was there *Genetics and the Origin of Species* in three separate editions within fifteen years, but Dobzhansky showed how one could go on modifying and improving one's own position: how new problems are to be welcomed, opening up new vistas of research. First, there was all of the excitement over the findings about chromosomal variations in *Drosophila* and the significance for our understanding of the working and scope of natural selection. Later, when the molecular revolution was upon biology, Dobzhansky's students (Richard Lewontin, in particular) were able to exploit it, seeing how it too is a friend of the evolutionist, making more problems (puzzles in the Kuhnian sense) than there had been previously. This is the stuff of real, professional, classy science. And Dobzhansky celebrated with the *Genetics of the Evolutionary Process* (1970), de facto, another edition of *Genetics and the Origin of Species*.

What about the nonepistemic front? At one level certainly the synthetic theorists were very different from Huxley in their attitude toward the open incursion of cultural values into the best kind of science. All told, Dobzhansky spent over ten years in Morgan's laboratory, a place which was contemptuous of social values intruding into serious professional science. For all of his Christian beliefs, Dobzhansky learned the rules. Consider the worrying question of deleterious variation in populations. Dobzhansky wrote (1937, 126):

> It is not an easy matter to evaluate the significance of the accumulation of germinal changes in the population genotypes. Judged superficially, a progressive saturation of the germ plasm of a species with mutant genes a majority of which are deleterious in their effects is a destructive process, a sort of deterioration of the genotype which threatens the very existence of the species and can finally lead only to its extinction.

Perhaps, suggested Dobzhansky, things are not quite so bad as they seem at first sight. Perhaps (arguing now in a way that was to flower with the balance hypothesis), the new mutations give rise to a reserve, held in hand in the event of change and new demands. Species have a kind of flexibility or plasticity, making them able to respond to needs. Although, added Dobzhansky quickly, whether or not this be true, it is inappropriate in a strictly scientific discussion to comment on the moral or religious implications: "A species perfectly adapted to its environment may be destroyed by a change in the latter if no hereditary variability is available in the hour of need. Evolutionary plasticity can be purchased only at the ruthlessly dear price of continuously sacrificing some individuals to death from unfavourable mutations. Bemoaning this imperfection of nature has, however, no place in a scientific treatment of this subject" (127). There is a time and a place for everything, and the time and place for culture and values lies not in science of the first rank, and especially not in science that one is trying to upgrade to the first rank.

We have seen much evidence that cultural values were overt in the thinking of Dobzhansky and his friends. To service his two sets of values, epistemic and nonepistemic, Dobzhansky's gambit (and that followed by others) was to write two sets of books and articles. The first was professional science. It might have been needed to support cultural views, but it was supposedly without culture or values itself. *Genetics and the Origin of Species* (all three editions and the retitled revision) fell into this category. The second set, by contrast, was explicitly written for the popular or general audience: mathematics removed (an easy job!); value discussions (especially progress) included; and a disclaimer at the beginning that this is written for the nonscientific reader. Epistemic factors would not be expelled entirely and certainly not violated—no pseudo-science wanted here—but the books were not to exclude cultural values. The aim would be popular science, as the titles show: *The Meaning of Evolution* (Simpson); *The Biological Basis of Human Freedom* (Dobzhansky); *The Biology of Ultimate Concern* (Dobzhansky); *The Basis of Progressive Evolution* (Stebbins). In this way, Dobzhansky and friends felt they could be true both to their science and their culture.

It is important to be aware of the existence of these books of the second series. It would surely be the claim of the objectivist that too often the subjectivist/constructivist unfairly concentrates on these, mistaking them for books of the first series. There is good reason why culture can be found in science, the objectivist would say, because science at this popular level positively invites cultural underpinnings and interpretations. But whether you agree with Simpson about evolution and ethics and the related virtues of democracy, and however you interpret the exact relationship between his biology and his conclusions in these sorts of discussions, is beside the point. The question of real concern must be less about the second series of popular works and more about the books and related articles of the first series.

As with the English evolutionists of this time—those who were so critical of Julian Huxley—we have at work here a metavalue that goes back at least to Cuvier, a value of the internal culture of science itself, namely, that of keeping science distinct from culture and hence nonepistemic-value-free. Yet, whatever the motivation, the fact remains that the synthetic theorists thought that the best science is of worth precisely because it is free of cultural values and therefore supposedly objective. Which means that we should ask: How successful were Dobzhansky and co-workers at keeping nonepistemic values out of their science? Did they truly succeed in their aims?

The conclusion has to be that at most they were only partially successful. This was true of Dobzhansky, and it was true of the others. Simpson, for instance, was most careful not to talk about progress in his professional books, *Tempo and Mode in Evolution* and the rewritten and retitled edition, *Major Features of Evolution* (1953). But he was a progressionist, as he makes very clear in his popular work, especially *The Meaning of Evolution*. And sometimes the enthusiasm for this notion seeps through to the professional work—no doubt a reason why Simpson felt able to revert to progressionist justifications for the ethics. Certainly, he saw niches waiting for organisms to occupy them and thought that thereby life climbs ever higher. Mammals after reptiles and so forth. Likewise, the infiltration of the cultural into the best science is true of the work of the others. Even Stebbins (1950), the botanist, talks about the higher plants and how evolution pushes upward to produce them. The late appearance and great success of the flowering plants (angiosperms) is apparently no accident.

How could this cultural-value infiltration occur? Quite simply. As we know, in this group, central to everybody's thinking—central to their professional, mature, science thinking—was the shifting balance hypothesis of Sewall Wright. Even Mayr, who in later years put distance between himself and Wright, based his original picture of the evolutionary process on the hypothesis. But a key component to the balance hypothesis, realized and seized upon by everyone, is the metaphor of an adaptive landscape. Organisms sit upon the peaks of hills, surrounded by valleys of low fitness; thanks to drift, they fan out down the sides of the hills; and every now and then some small population manages to cross a valley and then shoots up the other side. And with this hypothesis comes progress. Overall, given enough time, organisms will climb to ever-higher peaks—not just relatively, but absolutely.

Cultural value is built right in, although we must take care in understanding the exact relationship between the landscape metaphor and progress. I stress again that in itself the metaphor is not necessarily progressionist. One could think of the landscape as a choppy sea rather than something forged in granite. In which case there would be no real progress, for as soon as one climbed a peak it would collapse beneath one. Indeed, the possibility of this nonprogressionist interpretation of the landscape is vital to the theory. If progress had been an intrinsic part of the metaphor, no one would have touched it, because it would be too blatantly nonepistemic. However, the landscape metaphor does lend itself to—it invites—a progressionist reading. And so, having convinced themselves and most of their readers that they were not committed to necessary progress, the synthetic theorists tended immediately to interpret the landscape in a progressionist manner, as something which really does pave the way to genuine advance.

No doubt, similar arguments could be made in the cases of other nonepistemic or cultural values cherished by the synthetic theorists. But enough has been said to make the point. These values did occur in the professional or mature science, and they were important. However, one must qualify things by saying that they did not occur overtly. For that, one had to look to the popular science.

So what do we say in conclusion about the status of Dobzhansky's work? What was the standing of the synthetic theory as science? Deliberately, the synthetic theorists had set out to up-grade evolutionary studies, although not at all because they had ceased to think that cultural values were important. Indeed, the very contrary. They were motivated by urges very similar to the generations before them. Dobzhansky and Simpson were ardent progressionists, and Stebbins and Mayr still are. In the thinking of this group, social progress is possible and is a good thing, and the biological world mirrors this in being itself progressionist: from simple to complex, from controlled to controlling, from monad to man.

However, the synthetic theorists were driven by the desire to gain respect for their work as evolutionists and for the opportunity to work full time on evolutionary problems. They were under the spell of a metavalue, in the sense of something about rather than within science. The theorists wanted to move out of the museums and into the universities and to have all of the privileges and benefits of real researchers. They wanted their science to advance to the point where objectivity is a realizable aim.

I do not mean that these people were any less genuine in what they did, at least in the actions they performed and preached, than were the physicists, chemists, and mathematicians of their day. Sociologically, the synthetic theorists initiated the societies, journals, positions, grants, students needed to make a functioning science. Epistemologically, the synthetic theorists insisted on work which was empirical, experimental, quantifiable, as predictive as possible, consistent, and coherent. When they were not able to do things themselves—especially when it came to mathematics—they sought out the help of those who could, and they saw that their students would be trained above their own deficiencies.

However, they did what they did not just to gain a better understanding of the natural world but for the nonepistemic value of professional respect. They promoted epistemic values to further the nonepistemic goal of being considered mature and professional scientists.

Into the Present Era

One cannot say truly that the success of the synthetic theorists was absolute. Their timing was not the best (although when is the timing ever the best?). The 1950s and even before saw the dizzying rise of biochemistry and then molecular biology, offering masses of challenges to be overcome and (thanks to grandiose hopes from medicine to agriculture) with huge amounts of financial and institutional support. Evolutionists and others who studied whole organisms really did have to fight for a place in the sun. And it was not always easy to mount the fight, especially when one was asked about the practical payoffs. Nuclear testing can only go so far. One smart move made by the evolutionists in the 1960s was to get on the bandwagon with the trendy ecologists. But the money worries were there and indeed remain there to this day.

Finally, evolutionists had to confront the legacy of the past—the not-so-very-far past, as we have seen. Evolutionary theory had the odor of the philosophical about it, using "philosophical" in the pejorative sense reserved by scientists for the obsessions of the aged and the second-rate. People thought of evolutionary theory as something more than (and hence also less than) real science; and, for all the supposed separation between evolution and culture, the antics of someone like Dobzhansky, with his enthusiasm for Teilhard, did not help. Indeed, whatever the desires of the full-time evolutionists, others were quite content to see evolution keep to its old role. They wanted a "religion without revelation" (to use the title of one of Huxley's books), and they were happy to have a field other than their own provide it. Evolution had its place, and not everyone wanted it changed.

Yet, evolution did cross the divide. It was no longer just a dressed-up excuse for progress and related values carried over from the nineteenth century. It did become more than a popular science. It did get professionalized. Even though cultural values continued to play a significant role, the place where they were to be found overtly was in the popular writings (which were abundant) of the theorists. When (if) these values seeped back up into the professional, mature science, they were covert. Perhaps indeed they were in some sense eliminable, especially if put under threat.

More on this later. Now, let us move the clock forward to the leading evolutionists of our own day and see what they make of their science.

Karl Popper

Thomas Kuhn

Erasmus Darwin

Charles Darwin

Julian Huxley

Theodosius Dobzhansky

Richard Dawkins

Stephen Jay Gould

Richard Lewontin

Edward O. Wilson

Geoffrey Parker

Jack Sepkoski

6

✌

RICHARD DAWKINS

Burying the Watchmaker

In 1962 William Hamilton was very lonely. A graduate student in London, with little money, he would spend long days in Kew Gardens, and at night, when the libraries were closed, he would seek the proximity of other people in the halls of Waterloo Railway Station. To the disappointment of those who would engage him in conversation, anonymous sex was farthest from his mind. He was interested only in family relationships, for he was then trying to work through his ideas on how the forces of evolution could direct behavior socially toward others, especially those who share one's genetic heritage (Hamilton 1996). These were the concerns as Darwinism entered its second century.

Evolution is very much an idea of our day. It is always good for a television program or a dispute in the newspaper or—thanks to the dinosaurs—a high-tech movie packing in the millions who come to be amused, amazed, and terrified. Charles Darwin would be overwhelmed at the sheer ubiquity of his idea in the popular consciousness. Certainly one major factor accounting for this high profile is the attack on evolution by evangelical Christians (Ruse 1988). Nothing excites or draws attention to an idea more than the claim that it is dangerous or immoral, and the fundamentalist attack surely includes this. Prominent critics make explicit their belief that evolutionism is the thin end of a wedge at the broad end of which are sexual promiscuity and intellectual flabbiness, as well as a grave threat to private property (Johnson 1995).

Drawing more positive attention to evolution has been the discovery of protohuman fossils in Africa. Publicized by such powerful media as *National Geographic,* the paleoanthropologists (students of human fossil history) have shown us fabulous facts about our past—for instance, that just three or four million years ago our ancestors were creatures half our height with ape-sized brains who yet walked around upright (Johanson and Edey 1981). And perhaps most striking of all are those finds and reconstructions of the fabulous monsters of yesterday—grotesque reptiles, apparently wiped out in a cataclysmic explosion when a comet or asteroid smashed into our earth and caused a nuclearlike winter. Who could remain unmoved by *Tyrannosaurus rex,* with those huge teeth and jaws? Even now one awakes from sleep in a cold sweat.

But it has not been just in the public domain that exciting things have happened in the past few years, since the Darwin-Mendel synthesis. The molecular revolution, although a threat to evolutionism's status, could not go ignored, and in fact its influence on the field, as we shall learn, proved highly advantageous. And then at more conventional biological levels evolution has advanced in both theory and empirical understanding—through studies in nature and in the laboratory. Most striking of all have been the developments by Hamilton and others on the biological science of behavior, especially social behavior—so-called sociobiology. The explosion of interest in the genetic basis of social behavior—despite (or perhaps because of) accompanying disputes, to which we shall turn—has been the major event on the evolutionary scene in the past three decades. What was for many years a quagmire to be avoided by all right-thinking evolutionists has now become the area most attractive to bright students and others determined to make their mark.

Enough by way of trailer for evolutionism today. My intent is not to provide a comprehensive survey of every last finding or hypothesis in the field. I certainly want to look at some of the major advances and ideas. But as importantly, I want a sense of the activities, the methods, the relationships of today's evolutionism. For this reason, I shall stay with individuals and try to capture their full life's work, not simply picking out isolated pieces to suit a particular philosophical thesis.

But as we turn to do this, it must be acknowledged that one consequence of the professionalization of science impedes my intent:

the relative role of the individual is diminished. It is not that people are any less talented or hard-working but that, in a professional science, teamwork becomes more important—as does rapid criticism and revision. A Charles Darwin working on his own for twenty years without publishing would be impossible today. Imagine if Watson and Crick had tried to delay their paper on the structure of DNA until 1973! Hence the people chosen for discussion in the chapters that follow are primarily icons representing countless unmentioned others, rather than paragons who are necessarily overwhelmingly significant in their own right.

With this qualification, I propose a tripartite approach toward an understanding of the status of modern evolutionary thought. In this and the next chapter, I will consider two evolutionists (Richard Dawkins and Stephen Jay Gould) who are at the forefront of the popular culture of evolutionism. Then in the following four chapters, I will consider evolutionists more for their contributions at the professional level—first, two leaders who are coming to the ends of their careers (Richard C. Lewontin and Edward O. Wilson) and then two men who just are now moving past midcareer, the kind of men who have done enough to attract our attention but who still have mud from the trenches on their boots (Geoffrey Parker and John J. Sepkoski). I am aware that our subjects frequently work collectively with others. This collaboration is very much of the nature of modern science. I assume simply that our authors take responsibility for what is written under their names.

Getting Down to the Genes

My story begins in Britain with Richard Dawkins, student of animal behavior, who has spent almost all of his career at Oxford University. He has authored a succession of wonderfully written works, including *The Blind Watchmaker* and, more recently, *River Out of Eden* and *Climbing Mount Improbable*. However, he is rightfully best known for his first book, through which he burst onto the national scene in 1976: *The Selfish Gene*. Its brilliance at capturing ideas persists. Even after twenty years, its fame is undiminished. A recent poll of British readers' assessments of the 100 most important books of this century listed only two science books, both authored by fellow Brits: Stephen Hawking's *A Brief History of Time* and Dawkins's *The Selfish Gene*. One should add that heading this list was Tolkein's *The Lord of the Rings*, some fifty-four places above *Doctor Zhivago*.

The key notion in *The Selfish Gene* (Dawkins 1976, 74–75) is that of an *evolutionarily stable strategy*—an application of game theory to biological behavior pioneered by the English biologist John Maynard Smith (1982):

> An evolutionarily stable strategy or ESS is defined as a strategy which, if most members of a population adopt it, cannot be bettered by an alternative strategy. It is a subtle and important idea. Another way of putting it is to say that the best strategy for an individual depends on what the majority of the population are doing. Since the rest of the population consists of individuals, each one trying to maximize his *own* success, the only strategy that persists will be one which, once evolved, cannot be bettered by any deviant individual. Following a major environmental change there may be a brief period of evolutionary instability, perhaps even oscillation in the population. But once an ESS is achieved it will stay: selection will penalize deviation from it.

Consider a population of birds, some of whom behave dovishly and some of whom behave hawkishly. The "doves" threaten, but run away if there is any real fighting. The "hawks" jump into a fight and retreat only when badly injured. Why is the population not all doves? Because genes to make you hawkish would be favored by natural selection, since a hawk would always win in a showdown with a dove. Why is a population not all hawks? Because genes to make you dovish would be favored by natural selection, since a hawk would be forever hurting itself, especially in showdowns with other hawks; a dove would escape unharmed, especially when competing with other doves. However, you can easily show that under certain ratios, given initial benefits and costs, a population will be stably balanced between a certain number of doves and a certain number of hawks—or, if the behavior is within one individual, stabilized when individuals show dovish behavior part of the time and hawkish behavior part of the time.

Note how important in this discussion has been an individual selectionist perspective, as opposed to one which makes the group the central unit of focus, very much in the tradition started by Charles Darwin himself (Ruse 1980). One builds group effects out of individual interests, rather than taking the group as the basic entity. Continuing in this vein, Dawkins fanned out to look at other aspects of the animal world, especially as they pertain to animal social behavior.

He pointed out, for instance, that natural selection means that in many animal species you are going to get tensions among closely related individuals, particularly between the generations. Parents "invest" time and resources in their offspring. From the parents' perspective, it is in their biological interests to raise as many healthy children as possible. There is little or no point in raising one individual, even a superindividual, if the other children wither and die. From the child's viewpoint, however, being a superindividual is rather attractive. So, thanks to individual interests, conflict will develop when the interests of the older child do not coincide with the interests of the mother and the other children. "A mother wants to stop suckling her present child so that she can prepare for the next one. The present child, on the other hand, does not want to be weaned yet, because, milk is a convenient trouble-free source of food, and he does not want to have to go out and work for his living" (138).

The "battle of the sexes" also got treatment from Dawkins. Males and females do not necessarily have the same biological interests. "If one parent can get away with investing less than his or her fair share of costly resources in each child, . . . he will be better off, since he will have more to spend on other children by other sexual partners, and so propagate more of his genes. Each partner can therefore be thought of as trying to exploit the other, trying to force the other one to invest more" (151). Males can reproduce with ease and then move on. But in many species—mammals, birds, and others—females get stuck, literally, with baby. Males are in competition for access to females. That tells you much about their nature. Females are choosing, carefully if they can, which males they will mate with—the ideal male gives something in return, which may be help with raising baby or which may be the kinds of genes that made him attractive in the first place. His sons, who are also the female's sons, will have these genes also. In this context,

Dawkins is led to speak of females playing either the "domestic bliss" strategy, where they try to coerce males into helping raise the offspring, or the "he man" strategy, where they just go with the sexiest partner, hoping to pass on such characteristics to their own sons.

Following the great success of *The Selfish Gene*, Dawkins has been spreading his net more widely. For the past twenty years, he has appointed himself as major spokesman for the Darwinian selectionist cause. *The Blind Watchmaker*, published a decade after *The Selfish Gene*, is a paean to adaptationism; it even goes so far as to argue that not only is Darwinism the true theory of evolution but that it is the only possible theory of evolution! "My argument will be that Darwinism is the only known theory that is in principle *capable* of explaining certain aspects of life. If I am right it means that, even if there were no actual evidence in favour of the Darwinian theory (there is, of course) we should still be justified in preferring it over all rival theories" (287).

Dawkins seems to be skating dangerously close to making Darwinism true by definition: natural selection is a necessary truth or a tautology or some such thing. But as one reads on, one recognizes a strong empirical component to Dawkins's claim. His argument is the traditional Darwinian argument: that the most significant fact about organic nature is its adaptive complexity, and Darwinism and only Darwinism can explain this. For instance, when it comes to saltationism—the claim that evolution could have come about through and only through large new variations (macro-mutations)—Dawkins sounds remarkably like certain arguments of Darwin himself: "Anybody who wants to argue that mutation, without selection, is the driving force of evolution, must explain how it comes about that mutations tend to be for the better. By what mysterious, built-in wisdom does the body choose to mutate in the direction of getting better, rather than getting worse" (305–306)? Normally, variations are deleterious, and large variations are very deleterious. This is an empirical fact, yet one entirely ignored or minimized by saltationism.

There is much more in Dawkins's discussion of evolution that is innovative and exciting. He brings his knowledge of computers to bear imaginatively on such traditional chestnuts as the claim that, if its only raw material is random mutation, selection itself could never produce adaptive complexity. Through the skillful use of computer models, Dawkins shows precisely how selection can do this. Beginning with quite random combinations and working with mechanisms that are totally blind, one can readily generate end results that are as ordered as one could possibly wish or dream. There is simply no need to invoke design or anything else. More strongly: any other invocation is futile. In the words of Dawkins's most recent book: "On one point . . . I insist. This is that wherever in nature there is a sufficiently powerful illusion of good design for some purpose, natural selection is the only known mechanism that can account for it" (Dawkins 1996, 223).

Values in Dawkins's Science

Richard Dawkins is explicit in making his pitch to the general reader. He does not sacrifice everything to the most rigorous and theoretical of scientific ends, such as striving to obtain precisely confirmed predictions. Moreover, much of what he writes is reporting on the ideas of others, trying to make concepts and theories clear to the lay person. His aim is understanding rather than breaking new ground theoretically or empirically. This is not to deny that, whatever the level of discourse, Dawkins's books are intended to be scientifically respectable—more than that. Indeed, he himself tells us that he prefers not "to make a clear separation between science and its 'popularization'": in providing us with new metaphors and new ways of seeing, his work "can in its own right make an original contribution to science" (Dawkins 1989, ix).

Be this as it may, Dawkins does not break epistemic norms. In *The Selfish Gene*, for instance, the central idea of the gene as the ultimate unit of selection is supposed to be internally coherent and consistent with what we know from other fields—molecular biology, for example. As Darwin knew but many have forgotten, an individual perspective on selection is not a matter of personal preference or (as you might think) simply a nonepistemic desire to make evolution's workings as self-centered as possible. A group perspective has internal problems: although in the long run everyone might benefit, it is difficult to see why (other than in exceptional circumstances) a short-term devotion to self should not be preferred by selection. Adaptations directed toward immediate personal benefit would seem to be fitter than adaptations benefiting others, even if down the road all would gain with, and only with, the latter. Unfortunately, selection is necessarily a short-term process, without forethought for the future.

Coherence and consistency are not the only epistemic norms structuring Dawkins's work. Take predictive fertility. Through the notion of an evolutionarily stable strategy, not only can one explain much behavior hitherto puzzling but also one can make new inferences and predictions and extensions into the unknown. All sorts of ways in which the theory might be extended and used to work away at problems are suggested, with the implication that students who go this way will almost probably be rewarded with stunning successes at the end. Consider, for example, the fact that monkey females who have lost their own child sometimes steal babies from others: "It seems to me a critical example which deserves some thorough research. We need to know how often it happens; what the average relatedness between adopter and child is likely to be; and what the attitude of the real mother of the child is—it is, after all, to her advantage that her child *should* be adopted; do mothers deliberately try to deceive naive young females into adopting their children?" (Dawkins 1976, 110).

What Dawkins does not offer here is any follow-through of this suggestion, presenting work which tests these suggestions against the real world, either in nature or through experiment. This is what one expects of professional science. As a popularizer, this is not part of Dawkins's mandate. But let us not underestimate his achievement. The intention—let us be fair, the successful intention—is to present work which acknowledges tough standards. The spoken claim is that work before that which Dawkins describes was flabby if not downright inadequate as science. If there are not that many confirmed predictions, neither are there contradictions with accepted science in other fields.

Moving on to the nonepistemic or cultural, do we find that this figures in Dawkins's work? It is there, explicitly and sometimes proudly. At the very least, when someone writes of a "he man" strategy and a "domestic bliss" strategy he is reflecting attitudes of his society: in this case, a male-dominated Oxford-college society. It may be protested that no values are intended here, but one doubts that the language is going to find full favor with feminists. Nor will the speculation that, reservations aside, "it is still possible that human males in general have a tendency towards promiscuity, and females a tendency towards monogamy, as we would predict on evolutionary grounds" (Dawkins 1976, 177).

A huge amount of hostility to religion is also characteristic of Dawkins's writings—unambiguously reflecting Dawkins's own values. Recently, this hostility has become so obsessional and so overt that one might truly say that today this value—blasting religious beliefs—is a major reason why Dawkins writes as he does on what he does. No doubt Dawkins would see himself as deriving (according to strict epistemic criteria) certain truths about the world and then applying them to problems of theology—specifically, proving designlike effects through natural selection and thus making unnecessary appeals to a Designer. However, one might be forgiven for suspecting that some of Dawkins's enthusiasm for natural selection is precisely that it supports his view of life as a "pitiless" process without ultimate meaning. It is precisely because Darwinism can so substitute for Christianity that Dawkins finds the theory attractive.

Consider, for instance, the following discussion (1995, 105):

> Cheetahs give every indication of being superbly designed for something, and it should be easy enough to reverse-engineer them and work out their utility function. They appear to be well designed to kill antelopes. The teeth, claws, eyes, nose, leg muscles, backbone and brain of a cheetah are all precisely what we should expect if God's purpose in designing cheetahs was to maximize deaths among antelopes. Conversely, if we reverse-engineer an antelope we find equally impressive evidence of design for precisely the opposite end: the survival of antelopes and starvation among cheetahs. It is as though cheetahs had been designed by one deity and antelopes by a rival deity. Alternatively, if there is only one Creator who made the tiger and the lamb, the cheetah and the gazelle, what is He playing at? Is he a sadist who enjoys spectator blood sports?

Then, with the questions posed, Dawkins draws his conclusion. In the cosmic scale of things, there is no meaning whatsoever: "The universe we observe has precisely the properties we should expect if there is, at bottom, no design, no purpose, no evil and no good, nothing but blind, pitiless indifference." Poetically, he ends with words that seem to relish the collapse of traditional theology: "DNA neither knows nor cares. DNA just is. And we dance to its music" (133).

The Secular Theology of Richard Dawkins

If not the religion of Christianity, what about the secular religion of progress? In fact, Dawkins is a little misleading on this question. Both in *The Blind Watchmaker* and in other writings he is careful to inform the reader that he is not about to overstep the limits to professional science. He warns against "earlier prejudices" and assures us that "there is nothing inherently progressive about evolution." Dawkins is, nevertheless, very interested in arms races and in the way that they make for a kind of comparative improvement. He writes of gazelles getting faster when faced with the threat from carnivores like cheetahs, and of the cheetahs in tandem getting faster when faced with the prospect of their food supply getting clean away. To this he adds: "In the world of nations on their shorter time scale, when two enemies each progressively improve their weaponry in response to the other side's improvements, we speak of an 'arms race'. The evolutionary analogy is close enough to justify borrowing the term, and I make no apology to my pompous colleagues who would purge our language of such illuminating images" (Dawkins 1986, 180–181).

From the comparative progress of an arms race we go to the absolute progress of the evolution of that superior intelligent being, *Homo sapiens*—a slide reflecting Dawkins's fascination with (and obvious approval of) computers. He points out that military arms races today have moved much more to the electronic sphere than they were previously, with weaponry advancing more along this line than in terms of crude methods of attack and defense. He then introduces a very controversial notion, formulated by the brain scientist Harry Jerison, of an Encephalization Quotient, which is a kind of IQ measure across species.

Apparently, as a general tendency in the fossil record, brains get bigger over the eons, which no doubt reflects a rise in the EQ of the possessors. This, we are told, is surely the result of an arms race between those who would eat and those liable to be eaten. "This is a particularly pleasing parallel with human armament races, since the brain is the on-board computer used by both carnivores and herbivores, and electronics is probably the most rapidly advancing element in human weapons technology today" (190). Dawkins points out that, on Jerison's measure, humans do better than any other species. We are warned not to make too much of this; nevertheless, "The EQ as measured is probably telling us *something* about how much 'computing power' an animal has in its head, over and above the irreducible minimum of computing power needed for the routine running of its large or small body" (189).

I think that by this stage even the skeptic will be starting to read the message between as well as on the lines. If one came away thinking that evolution is progressive and that natural selection is the power behind the throne, one would be thinking no more than what one had been told. On reading Dawkins's more recent writings, where he has appointed himself the spokesman for militant atheism as well as militant Darwinism, one might be tempted to link the two. Certainly, the collapse of the God-hypothesis leaves a vacuum to be filled: "Any Designer capable of constructing the dazzling array of living things would have to be intelligent and complicated beyond all imagining. And complicated is just another word for improbable—and therefore demanding of explanation." Unfortunately, this is a demand which cannot be satisfied by conventional religion. "Either your god is capable of designing worlds and doing all the other godlike things, in which case he *needs* an explanation in his own right. Or he is not, in which case he cannot *provide* an explanation" (Dawkins 1996, 77). There is no escape except through selection. One might well be forgiven for concluding that, whatever the status of Christianity, secular religion is alive and well today at Oxford University.

All of this is interesting. Yet, many will think that the big questions about cultural values remain to be asked. In particular: What about social values and Dawkins's "selfishness" vision of evolution? As background we can agree without argument that the vision goes directly back, via the *Origin*, to the eighteenth century and to the free enterprise, *laissez faire* capitalism worked out by political economists, notably Adam Smith. And we can agree that at the heart of this theory lies the claim that things work best when everyone is following his or her self-interest. Misguided attempts at charity only make things worse. Historically, this view of the world—as if everyone were a Scotsman on the make (as J. M. Barrie wittily put it)—was embedded in the theological assumption that God, the "Invisible Hand," stood behind our actions, maximizing benefits from our individual selfishness. It is this theory that Dawkins translates directly into the language of genetics (1976, 38–39):

> What are the properties which instantly mark a gene out as a "bad," short-lived one? There might be several such universal properties, but there is one which is particularly relevant to this book: at the gene level, altruism must be bad and selfishness good . . . Genes are competing directly with their alleles for survival, since their alleles in the gene pool are rivals for their slot on the chromosomes of future generations. Any gene which behaves in such a way as to increase its own survival chances in the gene pool at the expense of its alleles will, by definition, tautologously, tend to survive. The gene is the basic unit of selfishness.

With respect to the level at which natural selection is supposed to operate—in particular, with respect to Dawkins's Darwinian belief that selection always favors the individual's benefit over the group's benefit—there are epistemic factors (like the need for coherence) which lead him to argue as he does. The question now is whether there are also nonepistemic factors involved and, if so, what are they? It can of course be pointed out that in speaking of genes as "selfish" Dawkins is speaking metaphorically. It does not necessarily imply that the beings, the "survival machines" in which genes find themselves, are likewise selfish, in the literal sense of thinking only of Number One. However, it could have this implication, and no doubt this is the spirit in which many who have read Dawkins's writings have taken him.

To be honest, the true record of Dawkins himself is somewhat ambiguous. Since our interest is less in Dawkins as an individual and more as a representative of modern popular writing on evolution, we can afford to be charitable. In his animus toward religion, real nonepistemic values surely are showing through. This is a man who thinks religion a bad thing and tells us so. Nonepistemic values could also be found in his attitude toward women—although my suspicion is that Dawkins would claim that there truly are differences between males and females in sexual emotions. Perhaps the thinking on progress is likewise value-impregnated, especially given that so many in our society today do not regard belief in progress with much favor.

In the case of social and economic values, however, let us agree that nothing too much is being put forward. Certainly, whether or not Dawkins thinks or thought humans innately selfish, he is hardly saying that this selfishness is a good thing. He is not the biological equivalent of Gordon Gekko, the Michael Douglas character in the movie *Wall Street*, who praised greed both for its own sake and for its results. At most we are getting a sense of pride for candor and plain truthful speaking (as Dawkins sees it). There is the very human satisfaction in telling us "naught for our comfort."

Cultural Factors or Cultural Values?

Whether or not all that Dawkins claims is based on nonepistemic or cultural values, it is certainly based on nonepistemic or cultural *factors*. This is no less true of the ideas about selfishness than of the ideas about religion. To articulate his thinking, Dawkins uses elements of his culture, irrespective of whether he endorses them. We have all of these socioeconomic models in our culture about the workings of human nature; whether Dawkins himself thinks they reveal ends worth pursuing or not, the fact is that he uses them in his evolutionizing. They are vital to the pictures he paints. More than this: these cultural elements are crucial for Dawkins's attainment of the epistemic virtues of his theorizing. The selfish-gene theory comes from culture and leads to predictive possibilities and coherence and so forth. The work is a package deal.

I will say no more here on these matters, but they will reappear.

7

STEPHEN JAY GOULD

Speaking Out for Paleontology

Early in December of 1981, the federal courtroom in Little Rock, Arkansas, was packed. It was the first week of a trial brought on by the American Civil Liberties Union to challenge the constitutionality of a state law passed earlier that year. The law mandated "balanced treatment," in the publicly supported schools, between evolutionary ideas and so-called Creation Science, better known as the early chapters of Genesis taken absolutely literally (Ruse 1988). By the end of the third day, the case for the plaintiffs was going well. Theologians had testified that Christianity had long interpreted the Bible metaphorically; a philosopher (me!) had argued that Creation Science fails every criterion of demarcation between science and pseudo-science; and the scientists were pointing to error after error in the claims of the literalists.

Out at supper that night, everyone started to relax, and the wine flowed freely. Someone struck up a hymn—one of those stirring melodies from the Baptist South—ironically at first, but before long all were joining in with vigor. No voice was louder than that of Stephen Jay Gould: paleontologist, skeptic, Jew, New Yorker, Harvard professor, baseball fanatic. But then, no voice is ever louder than that of Steve Gould, which is a major reason why he is the best-known evolutionist in America today.

Punctuated Equilibria

Located in the Museum of Comparative Zoology at Harvard University, Stephen Jay Gould rivals Richard Dawkins in his fame as a popularizer of evolution. His *Ever Since Darwin,* a collection of essays published in 1977, was a bestseller, as have been several of his books since, especially *Wonderful Life,* his work on the long-lost organisms of the Burgess Shale, an outcrop in the Canadian Rockies. On the bestseller lists recently was *Full House: The Spread of Excellence from Plato to Darwin,* a work which shows how the nonexistence for nearly sixty years of .400 hitters in baseball has much to tell us about life's history.

Although this last work in particular was as popular as can be, Gould would be insulted and hurt were one to suggest that he is simply a writer and thinker of the public domain. He believes that he can straddle successfully the public and the professional, and would argue that as a professional evolutionist he is indeed making frontline advances. We shall have to consider this point. So for the moment, let us turn in a neutral fashion to Gould's work.

Gould began his career in the mid-1960s as a paleontologist specializing in the evolution of snails (Gould 1969). At this point, he was an orthodox Darwinian, who had written a review paper on problems of relative growth (a sometime interest of Julian Huxley, whose significance was acknowledged) showing how things considered nonadaptive can be fitted readily into a selectionist framework (Gould 1966). Soon, however, Gould was moving to make his own mark, most particularly with a fellow paleontologist, Niles Eldredge, in advocating a new paleontological theory of punctuated equilibria.

Together, Eldredge and Gould (1972) argued that the traditional synthetic theorist's vision of evolution as a smooth, gradual process, hidden only because of the incompleteness of the fossil record, is quite mistaken. The fossil record is not so very inadequate, and in any case there is theoretical reason to think that evolution will be jerky rather than smooth. "If new species arise very rapidly in small, peripherally isolated local populations, then the great expectation of insensibly graded fossil sequences is a chimera. A new species does not evolve in the area of its ancestors; it does not arise from the slow transformation of all its forebears." Hence, "The history of life is more adequately represented by a picture of 'punctuated equilibria' than by the notion of phyletic gradualism. The history of evolution is not one of stately unfolding, but a story of equilibria, disturbed only 'rarely' (i.e., rather often in the fullness of time) by rapid and episodic events of speciation" (84).

To make this case, Eldredge and Gould turned to the writings of Ernst Mayr (1959, 1963), who had proposed the so-called founder principle to explain speciation: a small group of organisms gets isolated; because of variation within the parent population, the group will have only a subselection of the total possible gene combinations; this will cause a rapid "shaking down" or "genetic revolution" among the members of the group as they learn to do with much less than the full complement; and so there will be rapid evolution to new forms.

I am not sure how far one would want to say that any of this was orthodoxly Darwinian. The founder principle seems to owe as much to Wright's notion of drift as to anything in the *Origin*. (Wright claimed it owed everything to his notion of drift.) Moreover, with the emphasis on speciation rather than adaptation, Eldredge and Gould were starting to think of life's histories less at the individual level and more at the group level—where overall patterns were to be understood through the dynamics of the arrival and disappearance of groups (what another evolutionist was to label species selection). But the general discussion was certainly placed in a Darwinian context, and at this point (1972) Gould was not setting himself up as a critic of the synthetic theory. He was merely arguing that people had not interpreted that theory properly when it came to macroevolutionary changes as shown by the fossil record.

However, as the 1970s rolled along, Gould started to get more and more uneasy with conventional Darwinism, especially with the assumption of ubiquitous adaptation. The main spur to skepticism undoubtedly was a massive reading program in the history of evolutionary thought that engaged Gould in preparation for a work he published in the same year (1977) as *Ever Since Darwin*. This book, *Ontogeny and Phylogeny*, part history, part science, argued that the much-despised connections between ontogeny (the course of development of an individual organism), especially in the embryonic stages, and phylogeny (the evolutionary development of a species) still have some worth; and in support he argued not only from evidence today but from the evidence of history. Since this history inevitably involved a great deal of German history, where the ontogeny/phylogeny analogy was taken most seriously, Gould immersed himself in that morphological tradition which had so infuriated Cuvier, *Naturphilosophie:* a holistic philosophy stressing that the most significant features of organic life are the isomorphisms which link organism to organism. Adaptation is in many cases secondary or non-existent, and unity of type or *Bauplan* (to use the German term for organic groundplans or archetypes) is primary (Russell 1916).

This led Gould to write numerous articles hostile to ubiquitous adaptationism, including a celebrated attack coauthored with the population geneticist Richard Lewontin, in which attention was drawn to the nonfunctional parts of the tops of church columns (which he called "spandrels"; see figure), and the moral was drawn for organisms (Gould and Lewontin 1979). By 1980 Gould was ready for an all-out assault on adaptationism, and he declared the synthetic theory of evolution to be effectively dead. Gould's version of punctuated equilibria (Eldredge has always remained more orthodoxly Darwinian) was now edging close to saltations—macro-mutations—for those crucial rapid changes in the course of evolutionary history (Gould 1980a).

Moves of this nature did not find favor with more conventional Darwinians, especially those working experimentally on rapidly reproducing organisms where natural selection is a vital tool. It was pointed out that saltations have no empirical foundation and that, on macro-scales, selection can do just about anything that you could want (Stebbins and Ayala 1981). Although hardly acknowledged formally, we see a consequent rapid retreat by Gould to a position that is certainly not inconsistent with Darwinian selection. However, it is not a retreat to the original position. Now Gould (1982a) was (and would still claim to be) offering an "expanded" Darwinism. Natural selection and adaptation are undoubtedly important when one is considering organisms in their day-to-day life and microevolution. But as one looks at more long-term matters, one sees that other factors, including brute chance, come increasingly into play.

Instead of the unilevel synthetic theory, one now has a hierarchical theory, that is to say something (like the Catholic Church) with differ-

The nonfunctional "spandrels" at the top of pillars in St. Mark's church in Venice. Technically, these are known as "pendentives," and they exist as a matter of architectural necessity, to keep the building up. That they can be used for decoration is a by-product. Gould and Lewontin (1979) argue that many Darwinians commit a fallacy akin to thinking the decoration the primary purpose. Many things in the organic world which may seem to have an immediate function (like decoration) are in fact nonadaptive by-products of the overall "architectural" constraints of a working organism.

ent levels. Down at the lowest level (the micro-level)—the level of immediate or short-term change—one has a Dobzhansky kind of evolution that is essentially a function of the genes under the control of natural selection and like processes. But then one has upper levels (the macro-levels), where one is thinking of evolution over long periods of time. Here one has different processes at work. This means that no one level (especially not the micro-level) is to be privileged (Gould 1982b). In the language often used in such cases, the upper levels cannot be "reduced" to the lower level, meaning (contra Dobzhansky) one cannot hope to explain away everything at the upper, bigger levels by expressing them in terms of the lower, smaller level.

Why is this? Well, one reason is that the actual physical architecture of organisms makes certain demands. For instance, if you have an organism with four limbs, you have to have a frame strong enough to carry them. And this means that not every change possible in theory is possible in practice. Changing one part of an organism may not be a biological option, given other parts of the organism, not to mention the molecular and physiological difficulties of simply achieving any end you may need or want. To put matters another way, we might say that biological necessity imposes certain "constraints" on possible routes of organic development.

And what this all means is this: although in some particular instance the pressure for change may build up, no immediate—certainly no general smooth—change is possible. The constraints rule it out. Then, as it were, in some cases the dam may break, the constraints may give way, and a rapid change may occur, switching organisms to radically new forms. However, since these changes are rare, one would not generally expect to find them at the microlevel. One would spot them only by turning to long-term studies, that is, to evolution at the macro-level.

Most importantly, one could not expect to explain such constraint-breaking changes purely in lower-level terms—natural selection and so forth. They are exceptional. Yet, although exceptional and inexplicable in lower-level terms, the changes one sees in the broader, macro perspective do have implications for our understanding at the lower level. Not only is reductionism challenged in the sense of the belief that everything at the upper level can be explained in terms of the lower level, but it is also challenged in the sense of the belief that the upper level can never have relevance for understanding causal mechanisms at the lower level.

Precisely because time does show that evolution can involve massive rapid (instantaneous or near-instantaneous) change, we should be very wary of claims about ubiquitous adaptationism. Perhaps the constraints of development mean that the new forms of organisms, their *Baupläne*, are not overwhelmingly functional. They are more accidental than anything else. Which means that adaptation is very much less widespread than is dreamt of in the Darwinian heaven. To assume otherwise—to assume that adaptation is general—is to indulge in excessive "Panglossianism" (so named after Voltaire's philosopher, Dr. Pangloss, who thought this the best of all possible worlds) or the building of "Just So" stories (so named after Rudyard Kipling's fantastical accounts of adaptations like the elephant's nose).

Much of Gould's writing in the past fifteen years or so has been concerned with fleshing out these claims and chipping away at the opposition—the ultra-Darwinian opposition, that is. One paper, for instance, was concerned with the shapes of certain species of shell, showing that atypical forms ("smokestack" shells) were due to constraints on growth rather than to the effects of Darwinian selection (1984, 191–192):

> Evolution is a balance between internal constraint and external pushing to determine whether or not, and how and when, any particular channel of development will be entered. Natural selection is one prominent mode of pushing, but most engendered consequences of any impulse may be complex, nonadaptive sequelae of rules in growth that define a channel. Most changes must then be prescribed by these channels, not by any particular effect of selection. Natural selection does not always determine the evolution of morphology; often it only pushes organisms down a preset, permitted path.

Another paper focused on the nonselective replacement of one form by another in the same ecological niche ("ships that pass in the night"; Gould and Calloway 1977). Yet another proposed language for nonadaptive features, "exaptations" (Gould and Vrba 1982), and a fourth discussed the nonadaptive patterns that one finds in the fossil record (Gould et al. 1977). Darwinism is not wrong, but it is a very limited part of the picture. The most recent writings continue in this mode. Although *Wonderful Life* has as its ostensive subject the fabulous Canadian finds of soft-bodied fossils dating from the Cambrian (over 500 million years ago), truly it is an attack on what Gould sees as a misconceived Darwinian picture of life's history.

Since, with work like this, we are at the point where it is very difficult to keep down the nonepistemic parts of Gould's writings, let us now drop all pretenses and turn directly to the question of values.

Values in Gould's Science

Of course Gould would think of himself as offering work of epistemic worth. Indeed, the chief public justification for the introduction of the theory of punctuated equilibria was that conventional interpretations of the fossil record—especially the excuses for gaps in terms of nonfinding or nondeposition of fossils—were simply ad hoc if not outrightly inconsistent with modern thinking about the evolutionary process. This was certainly the claim in the celebrated introductory paper of 1972: the founder principle is the way that knowledgeable people today think about speciation, and this should be the way that paleontologists think about speciation.

I take it also that, along with consistency, Gould would think of himself as offering work which is predictive (in the sense of telling you what to expect when you push your inquiries into the unknown) and which has other epistemic virtues as well. Presumably the hierarchical view of evolution's causes and processes is supposed to have some fertility function, leading to explanation of issues hitherto neglected or unappreciated. Certainly not less than Dawkins, Gould would think of his work as being or in line with good science. And whether we would give him everything he would claim—a point to be considered shortly—we can surely agree that the demands of the best kind of science are a serious aspect of Gould's work.

What of nonepistemic or cultural factors? Let us start with status. Although the study of fossils is the science ordinary people think of first when they think of evolution, in the professional world paleontology has low status indeed—far below the work of the fruit fly geneticist. All of those years when paleontology was found less in universities and more in museums, when entertaining or instructing the public was its chief function, when the significant theoretical occupation was making up hypothetical histories of life, have left their mark. A major factor motivating Gould is precisely that of upgrading the significance of his chosen subject. If his expanded Darwinism succeeds, then this means that paleontology must be taken seriously as a science. No longer will it be the puppy led by the geneticist master—a point once made very clearly by the title of a talk Gould (1983) gave before a conference of evolutionists drawn from all areas of inquiry: "Irrelevance, Submission and Partnership: The Changing Role of Paleontology in Darwin's Three Centennials, and a Modest Proposal for Macroevolution." (The three centennials were 1909, the birth of Darwin; 1959, the publication of the *Origin;* and 1982, the death of Darwin.)

I have been using the term metavalue to refer to an interest or norm which is about science rather than directly in it, one which shapes and justifies the content of the science rather than actually entering into the science as content itself. The desire to upgrade one's own branch of science is such a metavalue, and this particular desire shows just how important metavalues can be for a scientist. The original joint paper on punctuated equilibria was announced as a major conceptual advance. A year before, in 1971, Eldredge had published a paper on the subject, where the tone was of a minor correction in thinking about the fossil record. Now, with Gould on board, we were basically told to regard the new ideas as a paradigm shift.

And then, emboldened by his move, Gould went right on the offensive, suggesting that it was time for the geneticists to get their own house in order and to fit in with the findings of paleontology: a dramatic reversal of the usual epistemological roles. This bid for dominance was soon squelched, but still the claim was made that—thanks to hierarchy—one needs to consider paleontology as an equal partner with genetics. No longer could one (Dobzhansky-style) infer the macro from the micro. The logic of evolutionary reasoning is changed significantly. Hence, however you regard it, the desire to upgrade paleontology certainly played a role in Gould's formation and presentation of his science.

Move on next to the significance of Germanic culture. Gould's father was a Marxist, and he himself has boasted of this and of its significance for his scientific thinking, pointing to the analogy between the revolutionary picture seen by the Marxist and the likewise revolutionary picture endorsed and promoted by himself as a paleontologist (Gould and Eldredge 1977; Gould 1979). With fame and fortune has come a certain reluctance to be painted into so tight an ideological corner, and Gould has been rewriting history a little of late. But still, one can say that Gould and Marx draw on a common background—a background of and liking for Germanic idealism, where adaptation is played down and where isomorphisms and *Baupläne* are highlighted.

This was certainly reinforced by a number of other facts. Growing up in a secular Jewish family and educated in American public schools (where ostensibly church was separated from state), Gould did not have the steady diet of natural theology that someone like Dawkins would have had in British schools, especially British private schools. Living in New York, with a passion for baseball rather than natural history, Gould would not have spent his youth collecting butterflies from the meadows, while singing "All Things Bright and Beautiful" at school assembly. By his own proud admission, Gould's early encounters with biology consisted of looking at dinosaur skeletons in the American Museum of Natural History, and the collecting he did was of baseball cards.

Then remember that all of this was reinforced in the 1970s when Gould embarked on researching and writing his magisterial *Ontogeny and Phylogeny*. From thence forward Gould was as inclined, if not more inclined, to think from the *Naturphilosoph* paradigm as from the Darwinian one. And here certainly one feels that there was (and is) more than disinterested understanding. There is positive enthusiasm—pride in standing in line with the great German biologists of the past.

Finally, and most important, is the dispute which arose in the 1970s from the attempt of the sociobiologists to apply their ideas about the evolution of social behavior to our own species. Gould (1980b) was one of the major actors. Perhaps because of his Marxism, perhaps because of his Jewishness, he wanted no truck with direct applications of bio-logical theory to human social thought and action. Whatever the reasons, by the end of the 1970s Gould had become convinced that biological progressionism—Darwinian progressionism, at least—is part and parcel of the sociobiological program, and a major impediment to any kind of genuine social progress. He saw it as the justification for claims about biological differences between humans, with some (Anglo-Saxons) being held up as innately superior to others (Blacks, Jews, Native Americans). Particularly galling to Gould were the immigration laws passed earlier in the century in the United States, which made it very much harder for Jews to emigrate from Europe, and he was driven to write a whole book on the history of this subject—*The Mismeasure of Man*—showing how racial progressionism is no more than prejudice dressed up to look like science. This is "biological determinism" at its worst. "The concept of progress is a deep prejudice with an ancient pedigree . . . and a subtle power, even over those who would deny it explicitly" (Gould 1981, 159).

Of course, one might say that Gould was arguing against culturally infested science (or pseudo-science) but that his own nonprogressionist picture is simply a culture-free disinterested vision of objective reality. Interestingly, however, Gould himself eschews this path, arguing and agreeing that any discussion about these sorts of issues is bound to be culturally impregnated. "I criticize the myth that science itself is an objective enterprise, done properly only when scientists can shuck the constraints of their culture and view the world as it really is . . . I believe that science must be understood as a social phenomenon, a gutsy, human enterprise, not the work of robots programmed to collect pure information" (1981, 21).

By his own admission, Gould himself was promoting a cultural vision of the world, and the stage was now set for fifteen years of arguing against Darwinian progressionism. This culminates with his recent book, *Full House* (1996, 216):

> If one small and odd lineage of fishes had not evolved fins capable
> of bearing weight on land (though evolved for different reasons in
> lakes and seas), terrestrial vertebrates would never have arisen. If a

large extraterrestrial object—the ultimate random bolt from the blue—had not triggered the extinction of dinosaurs 65 million years ago, mammals would still be small creatures, confined to the nooks and crannies of a dinosaur's world, and incapable of evolving the larger size that brains big enough for self-consciousness require. If a small and tenuous population of protohumans had not survived a hundred slings and arrows of outrageous fortune (and potential extinction) on the savannas of Africa, then *Homo sapiens* would never have emerged to spread throughout the globe. We are glorious accidents of an unpredictable process with no drive to complexity, not the expected results of evolutionary principles that yearn to produce a creature capable of understanding the mode of its own necessary construction.

The irony of all this, of course, is that Gould does not argue as he does because he does not believe in social progress. Rather it is because he wants such progress! He thinks, however, that the only way to achieve it is through a denial of the racially impregnated evolutionary progressionist scenario. Such a belief is a carry-over from a discarded (or worthy-to-be-discarded) past.

Gould's Science in His Own Time

So much for the nonepistemic values and other cultural factors which shape Gould's thinking. We can certainly say that evolution as a vehicle for social values is alive and well and living as happily in Cambridge, Massachusetts, as it is in Oxford, England. As a final question in this chapter, we must now ask about the status of Gould's work. I have chosen him, like Dawkins, as a representative of popular science. What then of his aspirations as a professional scientist, one at the cutting edge of evolutionary theory? Do we simply maintain a discreet silence at this point? Surely not; Gould is respected by paleontologists and was recently made a member of the National Academy of Sciences. Far more than Dawkins, he publishes in professional journals and similar outlets—and generally one can tell when he is acting in his professional mode (the discussion of constraints in *Paleobiology*) and when in his popular mode (the critique of immigration laws in *The Mismeasure of Man*).

Nevertheless, one might predict (in light of the striving for status for evolutionary theory) some uneasiness among scientists about what Gould is doing. At a minimum one would expect that professional evolutionists would cordon off his popular work from his professional work and that they would be very uncomfortable about his tendency to blend the two, using the one (usually the popular) to promote the other (the professional). And since he so blatantly advertises punctuated equilibria as a means of gaining status for paleontology rather than strictly on epistemic grounds, one might hazard that this concept would fall on stony ground, especially outside paleontology.

All of these forecasts prove true. If one makes a count from the *Science Citation Index*, one finds virtually no professional interest in Gould's popular writings (see box). This is not neglect of the man as such, for you do find a great interest in some of his professional writings—the graduate student paper on relative growth, for instance. But although *Ever Since Darwin* may be a great thing to read on an airplane, it is apparently not something one uses and cites when one is preparing an article for *Evolution*.

If you look at what scientists have to say about Gould and his style, the comments are frequently scathing, especially about the way in which he blends professional and popular. Expectedly, those who emote most strongly on this matter are precisely those who have been at the forefront of trying to provide and promote a fully functioning professional evolutionism. Thus, Laurence Slobodkin, founder of a department of evolution and ecology at the State University of New York at Stony Brook, moans that Gould "violates certain rules of etiquette," failing to work for "clarity in the dual sense of expository simplicity and in making oneself transparent so that the empirical world is visible through the text but the peculiarities of the author are invisible" (Slobodkin 1988). In this context, it is interesting to note that the original Eldredge and Gould paper was published in a book collection of articles. Gould (who added all of the flamboyant material about revolutionary changes) rather bullied the editor into accepting it "as is." Eldredge's prior, more subdued article had been published in the prestigious and orthodox scientific journal *Evolution*, after going through the conventional refereeing process. One doubts that it could

Citations of Popular versus Professional Science

As representative of Gould's professional writings let us take "Allometry and Size in Ontogeny and Phylogeny" (1966) and *Ontogeny and Phylogeny* (1977a); as representative of his popular writings, *Ever Since Darwin* (1977b) and *The Mismeasure of Man* (1981). And using the *Science Citation Index,* let us simply tot up the references to these works in the professional literature. We can compare citations over five-year periods, so that we are not confused by length of time since first publication. Of course, the nature of these citations ranges from the purely perfunctory to detailed utilization, but random sampling suggests that the perfunctory tends more to the popular publications and utilization to the professional publications, underlining the message of the raw data that the professional scientific community makes much more use of Gould's professional writings than it does of his popular writings.

	65-69	70-74	75-79	80-84	85-89	90-94	Total
1966	15	53	83	147	181	149	628
1977a			43	259	308	348	958
1977b			3	18	21	16	58
1981				16	50	44	110

Gould is not peculiar in this respect. By comparison, let us consider four works by Edward O. Wilson (to be considered in detail in Chapter 9): the jointly authored (with Robert MacArthur) *The Theory of Island Biogeography* (1967); *The Insect Societies* (1971); *Sociobiology: The New Synthesis* (1975); and *On Human Nature* (1978). The first three of these are professional and the fourth (by the author's somewhat reluctant concession) is not.

	65-69	70-74	75-79	80-84	85-89	90-94	Total
1967	33	182	530	680	590	585	2600
1971		79	337	426	448	379	1669
1975			516	713	463	348	2040
1978			6	46	36	15	103

The story is exactly the same. (It is true that *On Human Nature*—and indeed *The Mismeasure of Man*—figure more in the *Social Sciences Citation Index* than do the other writings, but not so significantly as to change the general conclusion.)

have been written any other way and still be accepted in that flagship of professional evolutionism.

Finally, "acidic" is the mildest term for the tone of comments made about punctuated equilibria by biologists at the center of evolutionary thought. In the course of a review of a book by another author in *The New York Review of Books*, where emotional, value-laden comments are virtually mandatory, John Maynard Smith took out time explicitly to fulminate against Gould's thinking (1995, 46):

> Gould occupies a rather curious position, particularly on his side of the Atlantic. Because of the excellence of his essays, he has come to be seen by non-biologists as the preeminent evolutionary theorist. In contrast, the evolutionary biologists with whom I have discussed his work tend to see him as a man whose ideas are so confused as to be hardly worth bothering with, but as one who should not be publicly criticized because he is at least on our side against the creationists. All this would not matter, were it not that he is giving non-biologists a largely false picture of the state of evolutionary biology.

What makes this critique particularly striking is that Maynard Smith is himself a sometime Marxist and to this day shares many social concerns with Gould.

Gould has replied to this attack in bitter terms of betrayal, speaking of people like Maynard Smith (and Dawkins) as "Darwinian fundamentalists." Although apparently Maynard Smith "has written numerous articles, amounting to tens of thousands of words," about Gould's work, "always richly informed," now sadly he has fallen away under the evil spell of adaptationist fanaticism (Gould 1997, 37):

> He really ought to be asking himself why he has been bothering about my work so intensely, and for so many years. Why this dramatic change? Has he been caught up in apocalyptic ultra-Darwinian fervour? I am, in any case, saddened that his once genuinely impressive critical abilities seem to have become submerged within the simplistic dogmatism epitomized by Darwin's Dangerous Idea [i.e. all-powerful natural selection], a dogmatism that

threatens to compromise the true complexity, subtlety (and beauty) of evolutionary theory and the explanation of life's history.

You might think that it is mainly the British who are critical of Gould as a professional scientist, but this is not true. And, as with the popular work, professional evolutionists have shown their feelings with their feet—or rather, with their unwillingness to use Gould's paleontological work. Virtually nobody (including evolutionists) outside of the paleontological community builds on Gould's theory of punctuated equilibria. It is instructive to look at how Gould's work fares with his fellow evolutionists—not just passing references but actual employment of ideas (see boxes). The contrast is as stark as night and day—and one suspects that the reason is simple. Whatever he might imply to the contrary, Gould's work really does not yield a cornucopia of scientific

Punctuated Equilibria as Significant Science

First let us look at the raw data with respect to citations, taking as the significant pieces the original, "Tempo and Mode in Evolution" (Eldredge and Gould 1972); the follow-up, "Punctuated Equilibria: The Tempo and Mode of Evolution Reconsidered" (Gould and Eldredge 1977); the climactic "Is a New and General Theory of Evolution Emerging?" (Gould 1980); and a synthesizing piece, "The Meaning of Punctuated Equilibrium and Its Role in Validating a Hierarchical Approach to Macroevolution" (Gould 1982b).

	70-74	75-79	80-84	85-89	90-94	Total
1972	14	119	218	153	139	643
1977		27	178	139	91	435
1980			81	53	26	160
1982			18	39	16	73

These are respectable figures, although punctuated equilibria theory seems not to be in the category of MacArthur and Wilson's island biogeography or Wilson's sociobiology, and less than Gould's own *Ontogeny and Phylogeny*, for that matter—both absolutely and with respect to staying power. But let us dig more deeply, trying to see if the figures just reflect the general hype around punctuated equilibria or if it is an idea actually used by working evolutionists (see next box).

Punctuated Equilibria, Who Uses the Concept?

The house journal of those interested in the issues which interest Gould and where he himself often publishes is *Paleobiology*, and the major journal for all evolutionists is *Evolution*. Starting with 1975 (the year in which *Paleobiology* was founded), how many articles in these journals make reference to Gould at all? How many articles refer to punctuated equilibria (using the criterion of reference to at least one of the above four articles)? And (looking now at content) how many articles are in some sense favorable to punctuated equilibria?

Citations from *Paleobiology*

	A	B	C	D	E	F	G
1975–79	177	11	18	31	8	3	7
1980–84	216	7	42	35	10	10	22
1985–89	226	5	30	56	9	8	13
1990–94	194	4	9	51	5	3	1
Totals	813	27	99	173	32	24	43

Citations from *Evolution*

	A	B	C	D	E	F	G
1975–79	469	1	6	30	0	1	5
1980–84	613	1	22	35	6	5	11
1985–89	538	2	7	56	2	3	2
1990–94	752	2	15	60	2	1	12
Totals	2372	6	50	181	10	10	30

A = total number of articles; B = articles by Gould; C = articles referring to punctuated equilibria; D = other articles referring to Gould; E = positive responses to punctuated equilibria; F = negative responses; G = neutral responses; C-G refer to articles not by Gould.

These figures point strongly to the conclusion that although Gould certainly has high visibility as a professional scientist, punctuated equilibria is not a great professional success. In *Paleobiology*, excluding articles by Gould himself (a very respectable 27), 35 percent refer to something by Gould, but only 13 percent refer to punctuated equilibria and a mere 4 percent respond favorably. In *Evolution*, excluding articles by Gould himself (6), 9.8 percent refer to something by Gould, but only 2.1 percent to punctuated equilibria and a mere 0.4 percent respond favorably. The neutral and critical articles are also slight. These are only references; the numbers actually using the ideas are smaller. Note that over time, in both *Paleobiology* and *Evolution*, favorable interest in Gould's work (E) is declining.

benefits—the payoff you expect from full implementation of the epistemic norms of good science. He may not break such norms, he may bow in their direction, but he does not devote his energies to their full satisfaction and implementation. The average working evolutionist is no better off with Gould than without him. Simply trumpeting one's science (in this case, paleontology) is no substitute for actually demonstrating its virtues—and no amount of rhetoric, especially not in *The New York Review of Books,* can conceal this fact.

In the end, criticisms of Gould are really nothing personal. Every evolutionist reads the column "This View of Life" that Steve Gould writes for *Natural History,* and almost every evolutionist reads Gould's books, buys them as presents for relatives, and recommends them to students. And we are all honored and proud if we can claim friendship with the man. But essentially those parts of his science that are promoted primarily to the greater glory of Stephen Jay Gould or to his discipline of paleontology are passed by. Evolutionists would be no less indignant than your most hardline Popperian objectivist if you tried to build a case for the nonepistemic, cultural-value-impregnated nature of science on the basis of much of it.

8

⤫

RICHARD LEWONTIN

Adaptation and Its Discontents

In July of 1964, five young population biologists met together at the summer home of the ecologist Robert MacArthur in Marlboro, Vermont. Joining MacArthur were the entomologist Edward O. Wilson of Harvard University; Egbert Leigh, a mathematician interested in community structure; Richard Levins, a theoretical ecologist then at Chicago and later to join the Harvard School of Public Health; and the brilliant population geneticist Richard C. Lewontin, just then moving from Rochester to Chicago and later also to go to Harvard. Truly a cabal of young turks, intending to take over evolutionary studies from their elders, they plotted and planned future strategies, intellectual and social (Wilson 1994).

Nor were their hopes in vain or their energies wasted. In the decades to come, they really did change the face of evolutionism. It is the work of two of these men, then friends and later bitter enemies, which is the subject of this and the next chapter. First I shall treat of the contributions of Richard C. Lewontin and then of Edward O. Wilson. Their disputes will be part of my story, but the central theme turns on the seminal contributions to evolutionary thinking that each man has made in his own right.

Population Genetics

Richard Lewontin, born in 1929, was Dobzhansky's prize student, studying with the great evolutionist at Columbia after undergraduate years at Harvard. To the credit of both men, Lewontin's whole career as a professional scientist has been as apprentice taking over where the master left off. Lewontin was to do things in the molecular realm that Dobzhansky could only dream of; and in the field of formal theory also there could be no true comparison. Lewontin shows effortless skill with the most ferocious mathematics. But the program was set by the older man: variations in individuals, variations in populations. In Lewontin's words (Schiff and Lewontin 1986, xii):

> My first research, as a PhD student, concerned the way in which genetic differences between organisms were manifested in different environments. That work, and everything I have learned since, has taught me that the organism is not determined by its genes, although its different traits are undoubtedly influenced to varying degrees by its genetic constitution. Most of my mature research life has been devoted to the other biological question raised by theories of inherent inequality: the problem of how much genetic variation actually exists between individuals within a species.

In fact, the young Lewontin hardly worked on just one problem, and his publications cover a range of Dobzhansky-style topics: variation, dispersal, population numbers and density, and so forth. One much-admired paper is very revealing about Lewontin's interests and attitudes, especially about the extent to which, as a Dobzhansky student, he was within the Darwinian paradigm and the extent to which he was not. "The Adaptations of Populations to Varying Environments" (Lewontin 1957) does indeed touch on adaptation. However, the chief focus of the paper owes little to the *Origin* or its author, being rather on the concept of homeostasis—a non-Darwinian notion made much of in the 1930s by the Harvard physiologist Walter B. Cannon (1931). This concept goes in direct line back to Herbert Spencer and is obviously yet another manifestation of the far older idea of a "balance of nature."

In its updated form, homeostasis is concerned with the self-regulating equilibrium that organic physiologies display in order to survive in fluctuating environments. One is not surprised to find Lewontin actually using explicit Spencerian language in speaking of homeostasis as "the nature of a feedback system or a system of dynamic equilibrium" (Lewontin 1957, 396). Homeostasis found its way into modern evolutionary biology thanks to Dobzhansky's concern to acknowledge and explain the great genetic diversity in populations. Extending the notion from individuals to populations—"A homeostatic population is one which can so adjust its genotypic or phenotypic composition as to survive and reproduce in a variety of environments" (396)—Lewontin's concern was to analyze the various kinds of homeostasis, populational and individual. Naturally, of particular interest was the key Dobzhanskian claim that heterozygosis (having two different versions, or alleles, of a given gene—one from each parent—rather than two identical versions) is associated with improved biological fitness. At this point, and throughout the article, Lewontin showed what was to prove virtually a defining mark of his scientific writings: an ability to see objections and counter-examples and an unwillingness to take these lightly. He does agree that heterozygotes may generally be fitter than homozygotes, but not before worrying his way through a mound of conflicting data and theories—which Lewontin called "contradiction" (404).

Before concluding, Lewontin extended his discussion to that fascinating species *Homo sapiens*. Apparently, through "creative homeostasis," humans transcend or escape their biology (407):

> Creative homeostasis, whereby individuals alter the environment in order to fit it to their demand, is in part individual and in part populational in nature. The group activity exhibited by bees and other social insects and reaching its highest development in man, while depending upon cooperation among members of the population, allows each individual in the population to be fit in a variety of environments, rather than depending upon differential fitness of genotypes for adaptation of the populations as a whole. The quintessence of homeostasis is the human intellect; not only the quintessence but perhaps the culmination of its evolution. The immense adaptive range of man, far transcending that of any other species, is made possible by his power to alter his environment adaptively. As this power grows, and it is yet growing, the force of natural selection must diminish along with the necessity of genetic plasticity.

It is no great surprise that "The Adaptations of Populations to Varying Environments" was reprinted in an influential series devoted to the social sciences.

Darwinian or not, this is work of high quality. The breakthrough contributions, however, were a decade off. These came through collaboration with fellow Chicago zoology member Jack Hubby. Using insights gained from the advent of molecular biology—namely, the identification of the Mendelian units of heredity with DNA—the two men were among the first to devise a method of going right down to the structure of the gene, thus determining how much variation one finds in naturally occurring populations of organisms. The technique, called gel electrophoresis, seizes on the different electrostatic charges on the cell's building blocks (polypeptide chains of amino acids) and enables one to distinguish similarities and differences between members of the same or different species (Hubby and Lewontin 1966; Lewontin and Hubby 1966).

To put it mildly, the results were staggering. In the fruit fly *Drosophila pseudoobscura*, the two researchers found huge amounts of variation from one organism to the next. A third of the genes studied seemed to have variant forms (alleles). At any one place (locus) on the chromosome, one could expect to find heterozygosity in an average of 12 percent of the cases (Lewontin and Hubby 1966, 608). With findings like these, the way now seemed open to ask and answer some of the key questions that had puzzled evolutionists, most particularly (as one might expect from a Dobzhansky student) questions about the relative merits of the rival balance and classical hypotheses concerning the genetic structure of populations. It was precisely to this issue that Lewontin turned in his prestigious Jesup lectures at Columbia in 1969 and in the subsequent book, *The Genetic Basis of Evolutionary Change* (1974).

What is most striking about the book is its tone. One would surely expect it to be a triumphant fanfare for population genetics and for its new molecular methods, cutting through and resolving old issues and pointing the way forward to yet more exciting avenues of research. It is anything but this. There is the tale of advance, certainly. Yet even more there is frustration and pessimism, as Lewontin faces barriers and paradoxes every way he turns, and the successes seem less satisfying with every move, theoretical or experimental.

First, Lewontin lays out the challenge. Muller, the champion of the classical position, had argued that populations are essentially uniform, with selection purifying them of the occasional mutant gene, except where the gene proves advantageous and thus rapidly becomes the norm. Dobzhansky, the champion of the balance position, had argued that populations contain masses of variation, much of it retained by selection thanks to the adaptive superiority of heterozygotes. Traditional, that is nonmolecular, methods had been able to make little headway on this dispute. Every approach was crippled by distortions and unreasonable assumptions that made the results worthless. Now, however, new techniques promised to tear down barriers and throw the light of understanding on the darkest corners of ignorance. But, wonderful though the methods of gel electrophoresis may be, "The mother-lode has been tapped and facts in profusion have been poured into the hopper of this theory machine. And from the other end has issued—nothing" (189).

Why the negativism? Because the terms of debate have been shifted and the barriers have come right back up. The molecular evidence, from the Hubby and Lewontin papers on, show simply massive amounts of variation in every natural species that has been examined. One organism after another—fruit flies, crabs, mice, humans—reveals that many of the loci have alternative alleles. No species falls beneath the 25 percent mark, some go over 75 percent (117). The balance hypothesis wins decisively.

Yet not so fast! The balance hypothesis supposes that variation is held in populations by natural selection, because heterozygotes are fitter than either homozygote. This is something that is still to be proven. And the classical-hypothesis supporters strike back by arguing that there are technical and other difficulties in supposing the kind of forces needed to hold so much variation in balance. Although the variation itself cannot be denied, from the standpoint of natural selection it must therefore be neutral, that is, it exists in populations at the molecular (genotypic) level but it is imperceptible at the physical (phenotypic) level and thus is not exposed to the forces of selection. Hence, what we might call the neoclassical position argues that molecular drift must be the true explanation of population variation.

Is there, in the end, no way forward? Has the molecular revolution deceived us all with false promises? The final chapter of *The Genetic*

Basis of Evolutionary Change tries a new tack. Perhaps we need a new approach, one which is more holistic in the sense of moving up the hierarchy of existence. Just as gas theory cuts through its problems at the molecular level by treating of things at the macro-level (the gas laws, as opposed to detailed treatment of every molecule), so perhaps this is what we must do in population genetics. We must consider as one unit the whole genetic constitution (the genome) of the organism.

Genes do not exist in isolation, or in random grouping like beans in a bag. They come packaged with other genes, and how one gene performs is very much a function of how other genes perform. Hence, a coarser level of treatment is perhaps a truer description of reality. "Context and interaction are not simply second-order effects to be superimposed on a primary monadic analysis. Context and interaction are of the essence" (318).

Later Work

No sooner had Lewontin's book appeared than he was engulfed by an event which radically disrupted his career. This was the dispute over the biology of human nature, a dispute sparked by the publication in 1975 of Edward O. Wilson's *Sociobiology: The New Synthesis* (see Chapter 9 for discussion). Lewontin felt strongly that this work, by his now-colleague at Harvard, was a bad book, scientifically and socially. Always interested in philosophy, Lewontin plunged more and more into the metaphysics and ethics of evolutionary biology, as well as into writing works of a more popular scientific nature. His professional output has diminished from earlier days, although the laboratory with its many students has continued, and Lewontin himself has always kept up a degree of scientific research of the most technical and professional kind.

Lewontin's continued concern has been with variation. For all its virtues, gel electrophoresis is too crude to pick up many possible changes at the nucleic acid level; it works at the next level, the product of the acid. Yet new techniques have been devised to go down to the macro-molecules of heredity themselves, and Lewontin has taken advantage of them. In some respects, at least judging by a review he wrote in the mid-1980s, Lewontin is still pessimistic about the possibilities of finding the true forces acting on the genetics of populations. There are still formidable theoretical and methodological problems. In other respects, however, one has the impression that perhaps we can now move forward (Lewontin 1985, 96–97):

> For the first time, there is a real likelihood that population genet-ics will be able to rid itself of its preoccupation with assessing the amount of genic variation in natural populations and with under-standing the role of selective and nonselective forces. This is possible because developments in molecular biology, in particular in genetic engineering and DNA sequencing, have made possible a qualitative change in the nature of the data. By sequencing genes from natural populations, we will be able to follow particular phylogenies of genes and distinguish those identical by descent from those that carry the same coding substitution but are not closely related by ancestry in the recent past. In this way, selective identity will be distinguished from historical identity.

More work has continued, including an interesting recent paper (coauthored with a number of Italian researchers) on the genetics of colorectal cancer. But to complete the direct treatment of Lewontin's evolutionary biology, let us look briefly at other later work of a less technical nature. At the head of the list must be that paper coauthored with Gould, attacking extreme Darwinian adaptationism. "The Span-drels of San Marco and the Panglossian Paradigm: A Critique of the Adaptationist Programme" tears right into those who would see adap-tive function throughout the organic world. "Often, evolutionists use *consistency* with natural selection as the sole criterion and consider their work done when they concoct a plausible story [about the adaptive function of organic characteristics]. But plausible stories can always be told" (Gould and Lewontin 1979, 588).

What would our authors offer in the place of ultra-Darwinism? In some respects, much that everyone (especially Americans in the Dobzhansky school) would find familiar. Genetic drift, for example, could be a reason for the way organisms have evolved. In other respects, the authors reach back to earlier philosophies. Perhaps the claims of the *Naturphilosophen* about archetypes and *Baupläne* had some truth. Perhaps certain constraints on development mean that organisms must take the forms they take, irrespective of adaptive advantage or needs. "[The continental tradition] acknowledges conventional selection for superficial modifications of the *Bauplan*. It also denies that the adaptationist program (atomization plus optimizing selection on parts) can do much to explain *Baupläne* and the transitions between them" (594). Perhaps we should think likewise.

Other writings have not been quite this hostile to Darwinism, but they are from the same stable. *Human Diversity* (1982), an attractive work in the Scientific American Library, stresses throughout how wrong it would be to see all human attributes as simple results of selection maximizing adaptation through its action on random mutation. The main theme of the book is the extent to which human beings differ and how very much of their difference is to be found within groups as opposed to between groups. "Of all human genetic variation, 85% is between individual people within a nation or tribe." Putting matters another way, if only Africans survived a world holocaust, we would still have 93% of human variation. Indeed, "If the cataclysm were even more extreme and only the Xhosa people of the southern tip of Africa survived, the human species would still retain 80% of its genetic variation" (123).

Lewontin acknowledges that some of the differences between peoples may be adaptive. Body shape is a plausible candidate. Cutting down on surface area in cold climates is seemingly adaptive. "Typically, the Eskimo has a large, chunky torso and short limbs, whereas the Dinka of Africa is tall and thin with very long arms and legs" (128; see figure). Yet, even here Lewontin reserves full commitment: "Although these trends seem to make good sense, there is no actual demonstration that they subserve greater survival and reproduction."

More generally, Lewontin is withering about claims that human traits and personalities and abilities can be connected in any straightforward manner with biology in general and adaptive advantage in particular. One simply cannot separate out genetic factors and environmental factors in any simple manner. Indeed, genetic and environmental causes are "inseparable" (68). (This comes under a heading of "The Interpenetration of Genotype and Environment.") Lewontin has zero sympathy for claims about biological superiority, either within or between "races" (a notion which is itself highly suspect). The idea that some genes are

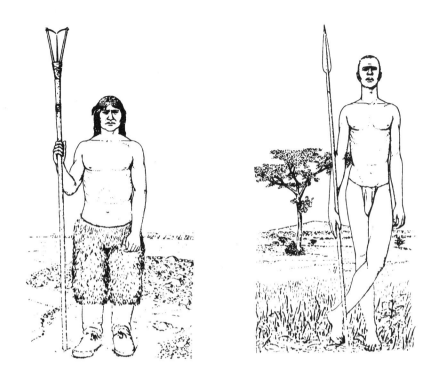

An Eskimo and a Nilotic Negro have very different ratios of body surface area to body volume (from Lewontin 1982, taken from Howells 1960).

better than others in the sense of determining the preferred charac-teristics of one person over another is ruled right out. Even talk of genetic "tendencies" is viewed with some disfavor (20).

Similar sorts of themes are found elsewhere, as in *Not in Our Genes: Biology, Ideology, and Human Nature,* a triauthored polemic against biological perspectives on humankind. They continue right through to one of Lewontin's most recent works, which started life as a series of radio talks on the Canadian Broadcasting Corporation. *Biology as Ideology: The Doctrine of DNA* is a sustained attack on what Lewontin sees as a pervasive bias in modern thinking, based on a perversion of biological thought (81):

> We are, in Richard Dawkins's metaphor, lumbering robots cre-ated by our DNA, body and mind. But the view that we are totally at the mercy of internal forces present within ourselves from birth is part of a deep ideological commitment that goes under the name

of *reductionism*. By reductionism we mean the belief that the world is broken up into tiny bits and pieces, each of which has its own properties and which combine together to make larger things. The individual makes society, for example, and society is nothing but the manifestation of the properties of individual human beings. Individual internal properties are the causes and the properties of the social whole are the effects of those causes. This individualistic view of the biological world is simply a reflection of the ideologies of the bourgeois revolutions of the eighteenth century that place the individual at the center of everything.

We are starting to leave the biological entirely and to get into philosophy. So let us now make the break cleanly, and turn to analysis of Lewontin's thinking.

Values in Lewontin's Science

No one could read Lewontin's work at any level and fail to recognize or be impressed by the fact that he, more than anyone we have met previously, takes seriously the epistemic side to science. From the beginning, his work is marked by attention to the demands of predictive accuracy, coherence and consistency, predictive fertility, and so forth. One might almost say that so seriously does he take these demands that they become, at times, almost paralyzing as Lewontin refuses to go out on a limb, beyond the epistemically secure. Certainly this is a feeling one can get from reading *The Genetic Basis of Evolutionary Change*, where almost every promising idea or path is shown quickly to be inconsistent with some other idea or path.

Nor is this dedication to the epistemic something which Lewontin shows by chance, as it were. It is a deep and integral part of his whole approach to science. At the beginning of *The Genetic Basis of Evolutionary Change*, Lewontin makes explicit his determination to play by the strictest of epistemic rules. One notion he introduces is that of "empirical sufficiency," which is something akin to what we are calling predictive accuracy. Lewontin argues that, all too often, population geneticists propose theories which are in themselves impossible to test. "If one simply cannot measure the state variables or the parameters with which the theory is constructed, or if their measurement is so laden with error that no discrimination between alternative hypotheses is possible, the theory becomes a vacuous exercise in formal logic that has no points of contact with the contingent world" (11–12). And even when accuracy is possible, too often its demands are not met. "The literature of population genetics is littered with estimates lacking standard errors and with methods for deciding between alternatives that have no sensitivity analyses or tests of hypotheses" (10).

The shadows on the past are that much deeper thanks to the gleaming epistemic virtues of the new method of gel electrophoresis. This is a textbook example of something which is fertile in the sense of opening up whole vistas of new problems, new solutions, new work. Quite literally, it transformed the way in which population geneticists approached their subject. Yet, pessimistic to the end, Lewontin titled a paper marking the quarter century of his seminal achievement "Electrophoresis in the Development of Evolutionary Genetics: Milestone or Millstone?" Nor was this an exercise in mock humility. For every milestone, Lewontin could think of a millstone. First, electrophoresis draws attention away from significant nonmolecular problems, for instance, those requiring selection experiments on aspects of morphology and physiology. These problems have not been solved: "They have only disappeared from our collective consciousness." Second, there is the complaint made in *The Genetic Basis of Evolutionary Change* that the present theory is not strong enough to account for the new facts about genetic variation. "So, ironically, the methods introduced to break the old impasse of evolutionary genetics has created a new and more frustrating impasse precisely because the data are so tantalizingly clear-cut and universal" (Lewontin 1991a, 661).

What now of the nonepistemic or cultural values? Lewontin more than anyone (well, perhaps Gould excepted) is given to accusing evolutionists of knowingly permitting nonepistemic values to contaminate their work. At the beginning of *The Genetic Basis of Evolutionary Change* we learn that the "balance school is strongly influenced by nineteenth-century optimism about evolution as being essentially progressive," whereas the classical school is "deeply pessimistic." Hence, "Genetic change can only be a change for the worse, and the function of natural selection must be to prevent degeneration by maintaining the type" (30).

Lewontin tells us that these two positions reflect different social policies. The classical supporters wanting eugenics, to preserve or retain the best. The balance supporters wanting diversity, and prepared to take steps to maintain it. "Neither view admits the possibility that genetic variation is irrelevant to the present and future structure of human institutions, that the unique feature of man's biological nature is that he is not constrained by it" (31). Later, we learn that the old idea of progress is in fact giving way to thoughts of equilibrium, again fueled by nonepistemic factors. Lewontin speaks of "sociopolitical convictions about stability that are deeply held and are characteristic of the present stage of social and political development of the West" (269). This is a change from the progressionism of the nineteenth century; but cultural factors still ride high.

As you can imagine, if Lewontin is prepared to say this sort of thing about the people with whom he associates, to whom he owes his whole research program, when it comes to explicit opponents—notably the sociobiologists with their Darwinian attitude toward humankind—he gets positively nasty. Racial, sexual, political conservatism—fascism even—you name it, he finds it (Lewontin 1977). Lewontin locates these sorts of faults in the ultra-adaptationists whom he and Gould criticize in their famous spandrels paper. But what really upsets Lewontin is the fact that these opponents are not simply influenced by the nonepistemic but, in his eyes, are prepared to sacrifice the epistemic in the process. Here is a man who has spent his whole life ardently emphasizing the highest scientific standards, and now his fellow evolutionists ride rough-shod over the boundaries. No wonder he is mad.

But while all of this no doubt explains in major part why Lewontin does not work on some of the problems that have so engaged others in recent years, we have not yet faced the question of his own values and his own work. Since he is himself a population geneticist and since he has argued strongly for various kinds of equilibria—remember the paper on homeostasis—it would not be unfair to suggest that he too was trying to find a social system in the world of his fruit flies, much as he suggests is the practice of others. Whether or not one agrees that progress and equilibrium are antithetical—I myself have claimed that they are both part of the (neo-)Spencerian picture—Lewontin too would seem to be part of the culture. Indeed, although the evolutionary biology in the 1950s when Lewontin began his career was far more selectionist than at most other times, one might suggest that ambivalence toward selection is simply part of the American way of life.

But, by Lewontin's own admission, there were other factors. One thing influenced his thinking right from the beginning, and readers of his most recent writings will know that it is still a major influence. Lewontin grew up a Jew in urban America, at the time when in Europe fascism and Nazism raged triumphant, ending in the horrific destruction of six million Jews (Schiff and Lewontin 1986, xiii):

> Just as theories of innate differences arise from political issues, so my own interest in those theories arises not merely from their biological content but from political considerations as well. As I was growing up, Fascism was spreading in Europe, and with it theories of racial superiority. The impact of the Nazi use of biological arguments to justify mass murders and sterilization was enormous on my generation of high school students. The political misuses of science, and particularly of biology, were uppermost in our consciousness as we studied genetics, evolution, and race. That consciousness has never left me, and it has daily sources of refreshment as I see, over and over again, claims of the biological superiority of one race, one sex, one class, one nation. I have a strong sense of the historical continuity of biological deterministic arguments at the same time that my professional mature research experience has shown me how poorly they are grounded in the nature of the physical world. I have had no choice, then, but to examine with the greatest possible care questions of what role, if any, biology plays in the structure of social inequality.

Even in a piece which appeared in 1997, in *The New York Review of Books,* the Nazi theme came bubbling through, with Lewontin worrying that genetic counseling was the first step to a Hitler-type eugenics.

Certainly this has all shaped Lewontin's interests. He did not invent the obsession with variation; it was already there in the Dobzhansky program. He did not share Dobzhansky's desire to fuel a progressionist process up to the Christian godhead. But Lewontin found the program and joined it, obviously with his own motives in mind. More than this, as the early paper on homeostasis shows, from the beginning Lewontin was pushing for ways to get humans out of the biological loop, to show how we are not part of the general determined picture.

No doubt connected here with his early personal background is Lewontin's ambivalent attitude toward adaptation. As a modern evolutionist he cannot do without the concept; but he sees how claims about adaptive advantage and (even more) adaptive disadvantage have been used by people like the National Socialists to justify cherishing some humans and regarding others—Jews especially—as vermin or nonhuman. And in the sociobiologists he sees a revival of this philosophy of selective human adaptive superiority. An oft-highlighted link between the Nazis and the sociobiologists is the ethologist Konrad Lorenz, who saw genes as destiny and who supported the Nazis in some of his writings (Lewontin 1977).

All of this is reinforced by Lewontin's early social and geographical location. He fits the Gould mold, rather than that of the English ultra-adaptationists. The gulf, bordering on disdain, shows: "The British school . . . carries on the genteel upper-middle-class tradition of fascination with snails and butterflies" (Lewontin 1974, 30). Then, as time has gone by, other factors have come to the fore. Notoriously and very publicly, the Lewontin of the past thirty years has been a Marxist: he has even coauthored a work provocatively titled *The Dialectical Biologist*, dedicated to Frederick Engels, "who got it wrong a lot of the time but who got it right where it counted." And certainly, whatever else, this philosophy was a factor in the structuring of *The Genetic Basis of Evolutionary Change*, adding to the rather paradoxical air—almost a love of contradiction for contradiction's sake—that puzzled many readers.

The Hegelian dialectic is explicit for those who would see. One starts with a thesis and counter-thesis—the balance and classical hypotheses—apparently in contradiction over the biological variation to be found in populations. This is resolved through a synthesis: electrophoretic studies reveal massive variation at a molecular level. But now again one gets a thesis and counter-thesis: neobalance and neoclassical hypotheses over whether this molecular variation does or does not escape the influence of selection. Again one has contradiction and again one tries for synthesis: now a holistic theory about the genome. And so presumably the story would go on.

In later writings Marxism can be found repeatedly running through Lewontin's writings. This is found especially in the writings on adaptation, where Lewontin not only promotes holistic, nonreductionistic themes (a particular bugbear is the selfish-gene theory) but tries to guide his thinking according to Engels's laws of dialectics, most particularly the law of the "interpenetration of opposites." Lewontin stresses again and again that an adaptation is not a static thing but something which creates and is in turn created by its environment—the one blending into the other, and the other blending into the one: "The environment is not a structure imposed on living beings from the outside but is in fact a creation of those beings. The environment is not an autonomous process but a reflection of the biology of the species" (Levins and Lewontin 1985, 99).

Recently, Lewontin has even been suggesting that we might need to get away from the very notion of adaptation. It is the wrong metaphor. We need a more dynamic picture like "constructionism": "An organism's genes, to the extent that they influence what that organism does in its behavior, physiology, and morphology, are at the same time helping to construct an environment. So, if genes change in evolution, the environment of the organism will change too" (Lewontin 1991b, 86).

Sorting Things Out

So much for values in the fascinating world of Richard Lewontin. With all of his explicit cultural concerns, one might conclude that he is truly no different from Gould. Far from being a paradigm of a professional scientist, he is a creature of the popular realm. At least, he had better be, else one will spend time fighting off the objections of objectivists who will claim today's evolutionism is intellectually flabby. Or one will be explaining to all who will listen why it is that evolutionists (like Lewontin's teacher, Dobzhansky) took seriously the metavalue that good, mature science is nonepistemic-value-free, yet evolutionists like Lewontin are at liberty to disregard the norm.

In fact, things are not nearly this bad, and one can draw a clear line between Lewontin and Gould. Start with the fact that Lewontin's scientific achievements do include work of the highest epistemic quality. If we are talking about predictive fertility, we are talking about one of science's true heroes. The gel electrophoresis work opened up enormous dimensions of inquiry, even if Lewontin himself bemoans that it has all become a millstone rather than a milestone. Punctuated equilibria theory is just not in this league. However else one may feel about the man and his work, one has to take seriously and respect Lewontin's positive achievements.

Second, note how much of the nonepistemic for Lewontin is operating at the meta-level rather than directly within the science. The Jewishness, for instance, inclines him into an interest in certain kinds of problems and avoidance of others. Moreover, remember how it manifests itself, not by making "friendly-toward-Jewishness" an operative value but by making the satisfaction of epistemic values more stringent. So the actual work that Lewontin produces is, if anything, more acceptable at the professional scientific level rather than less as a result of the nonepistemic values he holds. What we do not get is the positing of superior genes in Jews or anything like that—a kind of racial equivalent of the "aristogenes" that the paleontologist H. F. Osborn (1934) posited to explain why he and his class are superior to all others. Of course, one might say that Lewontin's values cut him off from avenues of good research. This seems especially so when he starts bashing adaptationism. But these are faults of omission, not commission. The work that is produced is epistemically sound.

But is this last claim entirely true? Surely, the Marxism influences Lewontin's science explicitly, promoting "contradictions" and pushing away from reductionistic Dawkins-style adaptationism and toward constructivism? Yet, most if not all of the writing of this ilk falls into the public (nonprofessional) realm. This is certainly true of the Marxist collection of essays, *The Dialectical Biologist,* and even more so of the lectures that started life as radio talks. However prestigious they may have been, they were certainly not (and were not intended to be) professional science. The writing on humankind appears mainly in a textbook—or rather in a popular book which could be used as a text—and much of the writing against human sociobiology was likewise for general consumption.

It is true that *The Genetic Basis of Evolutionary Change*, with its Marxist-influenced structure, would seem to be a counter-example. But I have noted already that many people felt slightly queasy about it. Certainly, many British scholars felt that the dismissal of their work without serious argument badly distorted the true state of evolutionary affairs—and was inappropriate, too, given that it was done on explicitly ideological grounds. Some were irate (Clarke 1974). In any case, although the philosophy is there for those who would read, many people back then (I was one) simply did not pick up the Marxist message! They (we) felt that Lewontin was being deliberately paradoxical and that the holism at the end was simply him blowing his own trumpet, after all else had failed.

Fourth and finally, as a professional scientist with the highest epistemic standards, Lewontin does not always take his own nonepistemic directives seriously. Consider the question of reductionism. This somewhat slippery notion has several meanings; but the sense we are using, and that which Lewontin himself uses in his critique of Dawkins, is that by directing inquiry to lower, smaller levels, one can understand things or events at upper, bigger levels—the macro is expressed in terms of the micro. Such reductionism is usually defended on epistemic grounds, namely, that its practice leads to greater predictive power and fertility and so forth (Ayala 1974).

Think of the effect of the Watson-Crick model of the double helix, a reductionistic advance on the Mendelian gene. Yet, the whole practice is opposed by Engels's law of quantity into quality, which supposes that at certain levels of development and aggregation one gets entities (wholes) which simply cannot be understood in terms of parts. To give an example which Engels gets from Hegel, water gets colder and colder (a quantitative temperature change), and then it changes qualitatively from one form to another as it freezes into ice. Reduction simply misses this sort of thing. One must be a "holist," thinking in terms of "hierarchies."

But for all the talk about the evils of reductionism and the virtues of holism, when it comes to his actual science, no one is more reductionistic than Richard C. Lewontin. Take the gel electrophoresis work.

There were problems about variation at the Mendelian level of the gene. No solutions were forthcoming, and so the whole problem was reduced to the level of molecules: What is the variation down at the molecular level of the gene? You might say that this in itself is hardly reductionistic, especially since the whole point of *The Genetic Basis of Evolutionary Change* was to argue that such a turn to the micro did *not* confer understanding at a higher level. But Lewontin has gone on digging to even lower (yet-more-micro) levels in hopes of understanding at higher levels.

Consider the paper on colorectal cancer (Presciuttini et al. 1993). It is a straightforward discussion about how certain genes lead to cancer of the bowel, with the certainty of early death unless the diagnosis is early and the surgery drastic. This is highly reductionistic thinking, in the sense that what goes on in a person's physical experiences is explained at the micro level, in terms of certain bodily molecules. As are prospective remedies. There is not even a suggestion that, in the hope of alleviating the problem, the afflicted carriers might contemplate a lifetime's diet of fresh vegetables and bran cereal (114):

> Our study supports the view that there is a genetic polymorphism in the general population, whose different alleles confer a different degree of protection against the loss (or the inactivation) of a major colorectal cancer gene, affecting in particular the probability per unit time of developing a colorectal cancer among the F[amilial] A[denomatous] P[olyposi]-affected subjects. The phenotypic difference caused by this hypothetical variation appears to be high, since the difference between the two modes of cancer mortality distribution is about ten years. We have no reason to suppose that the same effect is not present in the general population, so that we anticipate that most of the people affected by sporadic colon cancer are homozygotes or heterozygotes for the "worst" of the polymorphic alleles. In this regard, our results constitute independent evidence supporting the theory of a high-frequency dominant gene associated with increased susceptibility to colorectal adenomas and cancers.

I do not mean to deny entirely the influence of Marxism on Lewontin's professional work, although I regard the Marxism as a layer on top of older, deeper cultural concerns. But I do warn that, as always when scientists wax philosophical, it is well to approach with caution. And with this point ringing in our ears, let us move immediately to our second giant in the contemporary field of evolutionary studies, Edward O. Wilson.

9

⤫

E D W A R D   O .   W I L S O N

Southern Baptist Meets Charles Darwin

In the spring of 1958, Stanford University offered the young Harvard assistant professor Edward O. Wilson a tenured position. It turned out to be an offer that Wilson could refuse, because Harvard countered with a similar position, which he accepted. Everyone in Cambridge, Massachusetts, was happy, with the possible exception of fellow department-mate Jim Watson, who was still waiting for his offer. Watson was not used to coming in second, especially not to a whole-organism biologist. He is but one of many who have had to take their place in line, behind Ed Wilson (Wright 1987; Wilson 1994).

Like Lewontin, Edward O. Wilson was born in 1929. From the American Deep South, he moved to Harvard for graduate work, climbing up through the faculty ranks to the Frank B. Baird Jr. Professorship of Science and then to Pellegrino University Professor, a post from which he has just retired. An early interest in herpetology (the study of reptiles and amphibians) was deflected when, as a youth, he lost the sight of one eye through a fishing accident. Thenceforth, Wilson made the ants his special field of study, becoming their most distinguished student since his intellectual grandfather (teacher of his teacher), William Morton Wheeler.

In the early work, particularly of the 1950s and 1960s, Wilson worked at the interface between evolution and ecology. Ernst Mayr, a senior member of Wilson's department, was an important influence.

Indeed, essentially Mayr's problems were Wilson's problems: biogeographical distribution, speciation, adaptation to different environments, and more. Wilson even followed in Mayr's footsteps to New Guinea (Melanesia) for his first major exercise in field work. Gradually, however, the younger man's own interests started to come to the fore.

Biogeography

Wilson began with detailed studies of the ways in which ants invade new territories, creating new opportunities and taking space from old inhabitants (Wilson 1959, 1961). A hypothesis about how invaders start by colonizing marginal environments before moving to direct competition with already-established species led Wilson to turn his attention increasingly to some of those problems engaging Lewontin: the conditions and nature of group homeostasis or dynamic equilibrium. (The same language is used.) Wilson was trying to understand, in modern terms, what he now sees as the old problem of the balance of nature. If one species arrives and flourishes, does this not mean that an old species must leave or go extinct? Is there only so much ecological space available? "If a hundred species invade a certain ecological guild, say night-flying fruit eaters or orchid-pollinating bees, roughly a hundred comparable species will disappear, with many exceptions accruing to special places and times. The rule was reinforced in my mind by the discovery of a simple relation between the area of each of the Melanesian islands and the number of ant species found in it. The greater the area, the larger the number of species" (Wilson 1994, 216).

For all their overlapping of concerns, note what a very different world this is from Lewontin's. The interest in and knowledge of the genetic compositions of invading and retreating species are nonexistent. From the almost obsessive worrying of Lewontin, we turn to a sphere where the genetic background is a given, and the focus is simply and wholly on the organism itself. We are into the domain of the naturalist. Although Wilson has conducted many experiments in the laboratory—he keeps on hand a huge colony of leaf cutter ants—his real interest is in organisms as they find themselves in nature, responding to their fellows and to their environments.

The direction of Wilson's thinking led to a very profitable collaboration with the mathematically gifted Robert MacArthur and to the "equilibrium theory of insular zoogeography" (MacArthur and Wilson 1963, 1967). The two authors proposed models for the immigration and extinction of species on islands and showed the conditions—dependent on such things as island area and distance from the mainland—under which an equilibrium in overall species numbers would be expected. This was more an ecological hypothesis than directly evolutionary, in the sense of treating species in the short term without direct reference to internal biological change. But clearly, if one were thinking of the long term (which was precisely Wilson's starting point), the theory laid itself open to modification in appropriate ways.

In the years after the equilibrium theory was formulated, Wilson and his student Dan Simberloff performed a celebrated series of experiments in the Florida Keys (Wilson and Simberloff 1969; Simberloff and Wilson 1969). Through fumigation, they destroyed all of the insect life on selected islets and then measured the rates at which such islets were recolonized, comparing them with a number of untreated "control" islets. The result of the work was that the experimenters felt they had good evidence for the MacArthur-Wilson hypothesis: "Although the numbers of species on the control islands did not change significantly, the species composition varied considerably, implying that the number of species, S, approaches a dynamic equilibrium value $\check{S}$" (Simberloff and Wilson 1969, 285).

Sociobiology

Moving to the next phase of Wilson's activities, we arrive at the theory for which he is most famous—or notorious, depending on one's point of view. Working on social insects, Wilson could not have been indifferent to the new ideas pouring forth on the evolution of sociality. In fact, he was one of the first to recognize the power of William Hamilton's (1964a,b) thinking about what has come to be labeled kin selection: the concept that organisms can improve their own reproductive strength (their inclusive fitness) by aiding the reproduction of close relatives, who share many of the same genes. Following Hamilton in applying the concept to the ants, bees, and wasps (hymenoptera), which have a funny mating system where females are more closely related to their sisters than to their mothers, Wilson concluded with Hamilton that inclusive fitness explains the evolution of sterile workers, which spend their lives raising fertile sisters rather than daughters (see figure). Wilson speaks of Hamilton's insight as a "paradigm shift" and argues that Hamilton succeeded dramatically because "he went on to tell us something new about the real world in concrete measurable terms. He provided the tools for real, empirical advances in sociobiology" (Wilson 1994, 317).

Tools that Wilson was eager to use. The biogeographical work essentially complete, in the 1970s Wilson produced what he refers to as a "trilogy." First a book directly on the ants and other social insects, *The Insect Societies* (1971). Next came his magnum opus, a magisterial survey

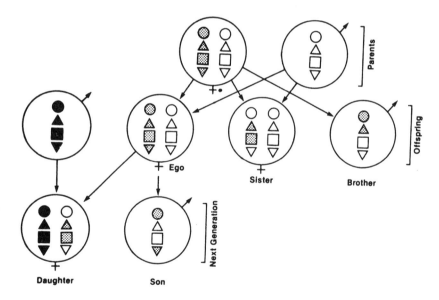

A diagrammatic representation of the genetic relationships in the hymenoptera. Females are diploid; males are haploid. Only females have fathers. Sisters have a 75 percent shared genetic relationship, whereas mothers and daughters have only a 50 percent shared genetic relationship. Kin selection therefore favors the raising of fertile sisters rather than fertile daughters. Males have no such special relationships and therefore do not form sterile worker castes (adapted from Maynard Smith 1978).

of all of the concepts and findings about the evolution of animal social behavior, up to and not excluding our own species: *Sociobiology: The New Synthesis* (1975; the echo of Huxley's earlier title was deliberate). Finally, as the decade drew to an end, a more popular work devoted exclusively to our species, *On Human Nature* (1978). For this last book Wilson won the first of two Pulitzer Prizes.

Sociobiology is an oversized, lavishly illustrated tome which announces at once that the author has moved on, dramatically, from the insects of the Florida Keys. If the title of the first chapter, "The Morality of the Gene," does not flag you, then the opening words surely will (3):

> Camus said that the only serious philosophical question is suicide. That is wrong even in the strict sense intended. The biologist, who is concerned with questions of physiology and evolutionary history, realizes that self-knowledge is constrained and shaped by the emotional control centers in the hypothalamus and limbic system of the brain. These centers flood our consciousness with all the emotions—hate, love, guilt, fear, and others—that are consulted by ethical philosophers who wish to intuit the standards of good and evil. What, we are then compelled to ask, made the hypothalamus and limbic system? They evolved by natural selection. That simple biological statement must be pursued to explain ethics and ethical philosophers, if not epistemology and epistemologists, at all depths.

In fact, after this dramatic opening, things do settle down rather. Having explained the way in which he sees sociobiology as an outgrowth of evolutionary ecology (see figure), Wilson next turns to a very detailed and comprehensive discussion of the causal factors behind social behavior. He covers the basic principles of evolution and genetics—Wilson may not use genetics in his own work, but he is acutely aware of its significance as a backing for all that he would claim—as well as the kinds of mechanisms which come into play in dealing with animal sociality. One topic that gets special attention, since it is based on work he himself did (in parallel with the work on biogeography), is chemical communication, especially between insects (231).

Then after discussion of other topics such as aggression, domi-

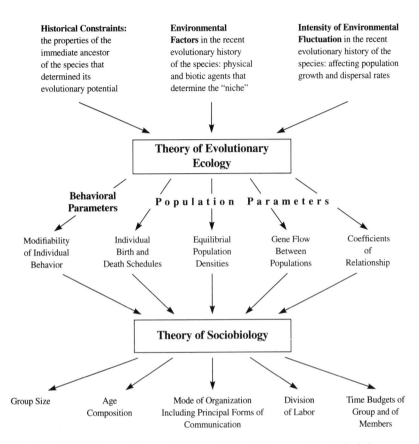

The place of sociobiology with respect to the rest of evolutionary and ecological theory, according to Sociobiology: The New Synthesis.

nance, caste systems, sexuality, parental care, and much more, Wilson is ready to turn to the really significant part of the book, a climb through the various social animals, beginning with colonial microorganisms and going all the way up to our own species, in a final chapter entitled "Man: From Sociobiology to Sociology." Nor should one think that the treatment of *Homo sapiens* was just added on as an afterthought. At the beginning of the discussion, we learn (379):

> To visualize the main features of social behavior in all organisms at once, from colonial jellyfish to man, is to encounter a paradox.

We should first note that social systems have originated repeatedly in one major group of organisms after another, achieving widely different degrees of specialization and complexity. Four groups occupy pinnacles high above the others: the colonial invertebrates, the social insects, the nonhuman mammals, and man. Each has basic qualities of social life unique to itself. Here, then, is the paradox. Although the sequence just given proceeds from unquestionably more primitive and older forms of life to more advanced and recent ones, the key properties of social existence, including cohesiveness, altruism, and cooperativeness, decline. It seems as though social evolution has slowed as the body plan of the individual organism became more elaborate.

It is in the pursuit of this paradox that Wilson structures his discussion, for apparently the "culminating mystery in all biology" is precisely how humans have been able to reverse the evolutionary trend away from social integration (382). In fact, we learn (from an argument consciously modeled on cybernetic thinking) that sometimes a kind of threshold is reached in (selection-fueled) evolution; this causes a sort of feedback; and thus a very rapid and significant form of evolution can take place. Applying this "autocatalytic" model to humankind, Wilson finds that such a process occurred first when we got up on our hind legs and coincidentally freed our hands for using tools and second when the brain exploded in size and mental evolution took over. This latter, "cultural evolution" (565–566) means that in an important sense we have transcended our biology; but, as Wilson makes very clear in later writing, it is only in a sense. He believes that, if certain moves or practices or beliefs prove particularly advantageous, biology is likely to track culture. Moreover, even as we stand today, apparently much that we think and do is subject to genetic control: at least as much as is produced by training or other environmental factors.

On Human Nature explores precisely how biology yet impinges on human consciousness and action. In the case of sexuality, for instance, we learn that male animals tend toward aggression, whereas females tend toward being "coy" and looking for males who will remain and help with child-rearing. "Human beings obey this biological principle faithfully" (125). Then there is help or altruism toward others: "Individual behavior, including seemingly altruistic acts bestowed on tribe and nation, are directed, sometimes very circuitously, toward the Darwinian advantage of the solitary human being and his closest relatives" (158–159). Finally, we have religion, a very important thing in the life of a Wilsonian human: "The highest forms of religious practice, when examined more closely, can be seen to confer biological advantage. Above all they congeal identity" (188). In other words, religion makes us feel that we belong to a group, and gives a meaning to our lives which reinforces our own self-interests.

Ants

By this point we are off the scale: off the scale of science, the critics would say. Yet, although claims like these kept Wilson at the heart of controversy for several years, even as the big sociobiology book was being published Wilson was continuing his insect studies. These intensified and led in 1990 to the huge volume, *The Ants,* coauthored with (then) Harvard colleague Bert Hölldobler, for which Wilson won his second Pulitzer Prize.

To understand the place of this work in Wilson's science, I will focus on a series of papers written by Wilson alone, around 1980, on the caste system and on how this plays out in the light of the so-called biological division of labor. The primary focus is on one genus of ants, the leaf cutter ants (*Atta*), famed of truth and fiction. They send out foragers from the nest, finding vegetation which they proceed to cut into small pieces to carry back to the nest. There, the vegetation is cut into even smaller pieces, treated with enzymes, and then used to grow a fungus, on which the young of the nest are fed. "The fungus-growing ants of the tribe Attini are of exceptional interest because, to cite the familiar metaphor, they alone among the ants have achieved the transition from a hunter-gatherer to an agricultural existence" (Wilson 1980a, 153). They are marked by having very different forms within the same nest: apart from the queen and the males, the female workers fall into distinct forms or castes, and the heart of Wilson's attention was the selective reasons behind such differences.

Wilson was prepared to accept that not every organic feature is necessarily at an adaptive peak. Indeed, as evidence that his approach was not *a priori* Panglossian, he would sometimes highlight features that are not. But his overall background assumption throughout the study is that life, in both physical form and behavior, is adaptive. It is produced by natural selection and is generally about as good ("optimized") as possible. This is as much the tool of Wilson's research as the hypothesis under examination.

What sorts of problems did Wilson tackle? First, there was the overall pattern to be found in the leaf cutters, the pattern of social integration. Here the discussion is more descriptive than causal and experimental, as Wilson strove to show that there is indeed a division of labor: members of one caste are more likely to perform tasks of one sort than members of other castes (who in turn have their appropriate tasks). In particular, the smaller workers spend their time right in the nest, working on the fungus fields ("gardening") or grooming the queen or tending to the young; the workers whose size falls into the mid range are out foraging and cutting up leaves and returning the pieces to the nest; and the very biggest workers (who can be up to a hundred times the size of the smallest) make up the soldier caste, whose concern is primarily or exclusively defense. "The elaborate caste system and division of labor that are the hallmark of the genus *Atta* are an essential part of the specialization on fresh vegetation. And, conversely, the utilization of fresh vegetation is the raison d'être of the caste system and division of labor" (Wilson 1980a, 150).

All of this variability is derived from a fairly simple base, biologically speaking. Natural selection does not design new forms for different castes. Rather, it works from a shared blueprint using flexibility in growth (allometry) and behavior to achieve its various ends. And this blueprint is the undifferentiated form of more primitive species, where all of the tasks are, or can be, done by all of the members. "Most of the monomorphic attines utilize decaying vegetation, insect remains, or insect excrement as substrates, in other words, materials ready made for fungal growth" (Wilson 1980a, 153). But to use fresh leaves, one needs specialists. Hence, the evolution of the caste system.

Is the colony well organized? Is it, to use Wilson's words, "as efficient in its basic operations as natural selection can make it, without some basic change in the ground plan of anatomy and behavior?" (1980b, 157). And how would one set about answering this question? "The ideal way to test the natural selection hypothesis and to estimate the degree of optimization is to first write a list of all conceivable optimization criteria, deduced *a priori* from a knowledge of the natural history of the species. The next step is to conduct experiments to determine which of the criteria has been most closely approached, and to what degree. Finally, with the results in hand, the theoretician can alter behavioral and anatomical parameters in simulations in order to judge whether the species is capable of still further optimization by genetic evolution. If the approach actually taken by the species cannot be significantly improved by the simulations, we are justified in concluding that the species has not only been shaped in this particular part of its repertory by natural selection but that it is actually on top of an adaptive peak" (Wilson 1980b, 158).

Take the question of evasion or defense. To find and cut up leaves, ants must leave the nest. This puts them at risk from predators and other natural dangers. Which members of the nest are actually used for this life-threatening work, and is this the most efficient use? The options run the gamut from using the smallest ants capable of cutting leaves, and trusting to evasion for protection, through using exclusively the best natural defenders, the biggest soldiers. In between these extremes is the option of using a mixed strategy of cutters and soldiers. Wilson, having determined that those out of the nest are on average significantly bigger than other ants in the group, ran a number of laboratory experiments using the "pseudomutant" strategy to test these various options. Each time, he removed all of the members of the external team except those of a predetermined size, and then checked on the efficiency in foraging and cutting of those that were left. This efficiency was compared to the cost of producing the external workers (judged as a function of body size, where producing larger workers is more costly than producing smaller ones). The range had to be limited because only workers above a certain size are able to cut up vegetation—and only workers above a certain larger size are capable of cutting up really tough vegetation (rhododendron leaves as opposed to rose petals).

The findings were unambiguous: "What *A. sexdens* has done is to commit the size classes that are energetically the most efficient, by both the criterion of the cost of construction of new workers . . . and the criterion of the cost of maintenance of workers" (Wilson 1980b, 164). Moreover, the ants have adapted in the direction of the ability to deal with a diet of uninterrupted tough vegetation, which, in nature, is much more likely to be encountered than soft vegetation. And finally, Wilson was able to show not only that the ants have adapted as they have but that they have adapted in such a way as to optimize the nest's collective behavior. "It sits atop an adaptive peak" (165).

More work of a similar kind followed, looking at the aspects of evolution in the leaf cutters and other genera (in particular *Pheidole*)—such questions as the extent to which different castes can take on new tasks in an emergency and how quickly a nest can rebound from a natural calamity. Throughout, the conclusion is that selection is an incredibly powerful mechanism behind an evolution leading to sophisticated optimized adaptation—if not in every instance—and that the division of labor in the social insects is a powerful instance of such adaptation.

Biodiversity

We come to the final phase of Wilson's work, perhaps more an inclination or a trend than an actually defined project. Wilson has been turning more and more from straight science to what we might call the philosophical or social. Apart from direct philosophy—for instance, two papers on ethics (coauthored with me)—an important mark of this phase was a little book, *Biophilia* (1984b), where Wilson started to float hypotheses about a biologically inherited love of nature possessed by all humans. Apparently we humans have evolved to such a point that we have a symbiotic relationship with nature: without it we would wither and die. In this we show our identity with the rest of organic creation. Indeed, even more so. We humans not only need nature physically to survive, we need it spiritually. A world of plastic would be deadly, literally.

From here, Wilson has slipped easily into ardent public enthusiasm for a long-held private obsession, which he calls biodiversity—life on earth, its complexity, and its interrelations. There is a paradox. "The biosphere, all organisms combined, makes up only about one part in ten billion of the earth's mass" (Wilson 1992, 35). How is it possible to do so much with so little? Although there is great loss at each level, the secret is the hierarchical nature of life. Life at the bottom level—green plants—takes in energy from the sun, albeit only 10 percent of the total energy which arrives on earth (1992, 36):

> The free energy is then sharply discounted as it passes through the food webs from one organism to the next: very roughly 10 percent passes to the caterpillars and other herbivores that eat the plants and bacteria, 10 percent of that (or 1 percent of the original) to the spiders and other low-level carnivores that eat the herbivores, 10 percent of the residue to the warblers and other middle-level carnivores that eat the low-level carnivores, and so on upward to the top carnivores, which are consumed by no one except parasites and scavengers.

Biodiversity is in a constant state of change. "Evolution on a large scale unfolds, like much of human history, as a succession of dynasties." Sometimes one form reaches to the top or is very successful, sometimes another form. Usually, decline is forever. However, "Once in a while, in a minority of groups, a lucky species hits upon a new biological trait that allows it to expand and radiate again, reanimating the cycle of dominance on behalf of its phylogenetic kin" (94). But overall there is a balance or equilibrium. "A limit to organic diversity exists so that when one group radiates into a part of the world, another group must retreat." Even though one might not want to speak of the balance of nature as a law of biology, this is a pretty standard rule. One group pushes another group from the center and into insignificance or extinction—or takes advantage of a gap which opens up. "The rise of the mammals after the fall of the dinosaurs is the textbook case, but examples exist among corals, mollusks, archosaur reptiles, ferns, conifers, and other organisms following the demise of their competitors in one of the major extinction spasms" (119–120).

Biodiversity is under continual threat, from natural disasters and the like. But fortunately it has the ability to spring back. The terrible

explosion on the island of Krakatau in 1883 was followed by recoloniza-
tion of what was left. "Today you can sail close by the islands without
guessing their violent history, unless Anak Krakatau happens to be
smoldering that day" (23). Even the great global extinctions of the past
were not enough to destroy biodiversity forever, although the time for
recovery was massive: major extinctions required 25 million years or
more (31).

Now, alas, we face the biggest extinction of them all: the human-
caused extinction. We are destroying species at a phenomenal rate; and,
most tragically, among the worst affected places are the rain forests and
jungles of the tropics—Brazil, for instance. We must do something
before it is too late. Else we will never see biodiversity again in our
lifetimes—or our children's children's lifetimes. This is more than a
pragmatic call. It is a spiritual warning. "Only in the last moment of
human history has the delusion arisen that people can flourish apart
from the rest of the living world." Would that we were more like
preliterate folk, who may not have understood the underlying principles
but who did grasp that "the right responses gave life and fulfilment, the
wrong ones sickness, hunger, and death" (349).

Values in Wilson's Science

With E. O. Wilson we have entered a very different epistemic world
from that of Richard Lewontin. Lewontin is ultra-cautious all the way,
rarely venturing beyond that which can be put to immediate test and
highly suspicious of sweeping hypotheses. By contrast, right from the
early days when he was speculating on the biogeographic patterns of
ants in the tropics, Wilson has been much given to such hypotheses.
And by their very nature, many things that he wants to claim cannot be
put to immediate test—at least, not ready experimental test. One has to
work indirectly, from natural experiments, for instance, such as
Krakatau.

Much of the tension between the two men—Lewontin and Wilson—is precisely over this difference in philosophy. Lewontin looks
down on those who do not have his own exacting standards. Wilson is
scornful of those who are not willing to take a chance—who fear the

"whiff of grapeshot" as one pushes out into the unknown. Daring conjectures risking rigorous refutations, to use Popperian language. We are not in completely different worlds. Lewontin and Wilson share the same epistemic norms, but, partly by personal nature and partly by the differences in material that they face, the two men put different emphases on the norms. Allometric growth has driven them apart, one might say.

Take the case of predictive accuracy. Wilson clearly cares about it. The work with Simberloff in the Florida Keys was guided by this norm, aimed at demonstrating such accuracy with respect to the MacArthur-Wilson theory of biogeography. And the feeling was that within limits the theory performed pretty well. One hardly has the precision that one might expect and demand in physics, but the theory came out considerably enhanced. Similar sorts of comments apply to the detailed work on ant castes and the division of labor. Almost self-consciously, Wilson spells out the range of possible results and why one might expect one part of the range to be satisfied rather than another. Then he sets to work and, as with biogeography, feels that his experiments are successful.

But much of Wilson's work is not really of this kind, or at least only tangentially. As often as not, he has a bright idea, some plausible arguments are offered, and then Wilson is off onto the next topic. Having been myself a coauthor with Wilson on papers about the evolution and nature of morality, I can speak with some authority in saying that this area does not lend itself to much in the way of predictive confirmation (Ruse and Wilson 1986). The same can be said of the criteria of coherence and consistency. Wilson is certainly not indifferent to the norms, but he will not let them stop him from pushing forward into the unknown—another example of the sweeping Wilsonian style that so deeply upsets critics like Lewontin. He has argued that many of Wilson's claims about humankind are either outrightly false or have been protected by so many safeguards as to make them unfalsifiable (Lewontin 1977). The claims that Wilson makes in *On Human Nature* about sexual differences are often cited as prime examples. Many critics have pointed out that no real empirical evidence for such claims exists and that much evidence goes against them. They can be maintained only by someone perversely blind to the facts.

In like manner, particularly blunt have been critiques of the work that Wilson produced after his trilogy on sociobiology, work he did of a formal nature with the Canadian physicist Charles Lumsden (Lumsden and Wilson 1981, 1983). Here, Wilson's interest was in the speed at which one might expect to find biological inclinations tracking cultural adaptations. His ideas have been severely faulted on the grounds that biological change could never occur as quickly as he claims, at least not without supposing selective forces of a magnitude never found in nature, certainly never found in human nature (Kitcher 1985). Lewontin argues strenuously that what Wilson wants is just not consistent with what he (Lewontin) has learned from his work in theory and in the laboratory (Segerstralle 1986). Epistemic norms are therefore being broken or dodged.

Move on from these uneasy areas to unificatory power and predictive fertility. Here, Wilson would want to shine—and indeed would congratulate himself on so shining. He would feel that sacrifices made in other epistemic directions are more than rewarded when it comes to these norms. Again and again Wilson tries for some grand hypothesis, synthesizing disparate areas of experience: biogeography, sociobiology, ant sociality, biodiversity. He is forever trying to build an overall picture. As far back as his early work on the ants of Melanesia, Wilson was trying "to synthesize certain information on the zoogeography, speciation patterns and gross ecology of a limited fauna." And it was this aim that shaped the product (Wilson 1961, 192):

> Three general attributes of success are recognized in the expanding Melanesian ant taxa: the acquisition of a significant ecological difference, which presumably reduces interspecific competition, the ability to penetrate the marginal habitats, and the ability to disperse across water gaps. It is suggested that the attributes are causally related in the sequence given. Success in the marginal habitats gives expanding species the advantage needed to encompass and progressively replace older resident taxa.

No less common than the grand unifying hypothesis is the model building and the proposals for whole new ways of looking at things—ways which hint at, if not promise, all sorts of fresh and exciting avenues of research. In a sense, the whole sociobiological synthesis is intended as an advertisement for new directions of scientific activity. The same is certainly true of individual parts of the synthesis. A good example is the so-called autocatalytic model of evolution proposed at the end of *Sociobiology* to explain human evolution. Not much by way of hard evidence is supplied in support, but new vistas of research are put on offer. By bringing in the idea of self-driven systems, with all sorts of feedback loops, Wilson is offering the would-be researcher in the area a new way of thinking about things.

In a less grandiose fashion but no less firmly, throughout his ant work Wilson is suggesting that his answers provide yet more questions for inquiry. In the *Pheidole*, for instance, there is the "surprising discovery" that in an emergency the large workers can very quickly switch to assume the role of the smaller workers (Wilson 1984a, 97–98). When he comes to the *Atta*, he speculates about the evolution of the genus. "Is it possible that one size group is closest to the ancestral, monomorphic attine species in anatomy and behavior?" If so, it "would be of interest to learn whether such a segment exists and to what extent it still bears the mark of its earlier, more generalist existence" (Wilson 1980a, 155). And then of course there is the need to go out from the laboratory and into nature. We must "analyze the resiliency of foraging workers under a wide range of natural conditions, including the abundance of the vegetation and its distance from the nest, as well as the effects of the size and amount of foraging activity of individual *Atta* colonies, before a definitive evaluation of the effectiveness of the resiliency can be made" (Wilson 1983, 53).

Critics who claim that Wilson is indifferent to the constraints of good science are simply mistaken. He is as caring about them in his way as is Lewontin in his way. The point is that the way of Wilson is not the way of Lewontin: the emphases are very different. But why? Turning to the cultural side of Wilson's thinking may yield rich dividends.

Edward O. Wilson came from the American South; he was brought up in the Depression years and then during the Second World War. No less than for Lewontin, childhood experiences left deep marks on the man. First there was religion. As a youngster, Wilson was "saved" and "born again" in Jesus Christ. For better or worse, this did not take for all time. After going to the University of Alabama, where a somewhat nerdish interest in natural history suddenly was both an asset and a ticket north, Wilson's religious fundamentalism fell away, to be replaced by the theory of evolution, which was regarded in the South with somewhat the same attitude that pagan Rome regarded the early Christians.

The post-Christian Wilson has always thought of his science in some way as a religion substitute and has used Darwinism not to banish faith but to find a more satisfying creed for the modern age (Wilson 1978, 201):

> The core of scientific materialism is the evolutionary epic. Let me repeat its minimum claims: that the laws of the physical sciences are consistent with those of the biological and social sciences and can be linked in chains of causal explanation; that life and mind have a physical basis; that the world as we know it has evolved from earlier worlds obedient to the same laws; and that the visible universe today is everywhere subject to these materialist explanations. The epic can be indefinitely strengthened up and down the line, but its most sweeping assertions cannot be proved with finality.

We are dealing with a "myth"; but, when all is said and done, "the evolutionary epic is probably the best myth we will ever have."

From such a man as this, one surely expects to find that childhood ideas and values of Christianity are incorporated into his science in some fashion. And this is true. Most obviously we have the ardent commitment to the natural theology of pure Darwinism, in a fundamentalist sense. Although he allows exceptions, Wilson's work is deeply adaptationist. For him, the living world is as exquisite in its design as it is for any true believer. Thus, his thinking is structured from the beginning by heavy cultural elements: adaptationism in general, and putative adaptations like the division of labor in particular.

The second mark is that of militarism. The South is and was a society where military prowess and valor are much prized. This too is reflected in Wilson's work. I do not now refer to his identification of such things as soldier castes in the ants. This is a universal metaphor for thinking about the ants. Rather, I have in mind the brave—some might say reckless—style of Wilson's science. By his own admission, this is a direct function of his youth and his own experience of its demands and pressures (Wilson 1994, 25):

> I have a special regard for altruism and devotion to duty, believing them virtues that exist independent of approval and validation. I am stirred by accounts of soldiers, policemen, and firemen who died in the line of duty. I can be brought to tears with embarrassing quickness by the solemn ceremonies honoring these heroes. The sight of the Iwo Jima and Vietnam Memorials pierces me for the witness they bear of men who gave so much, and who expected so little in life, and the strength ordinary people possess that held civilization together in dangerous times.

Explicitly, he links these feelings to his scientific method. Fearing he lacks true courage, he pushes himself to the limit. In grown life, "when I had ideas deemed provocative, I paraded them like a subaltern riding the regimental colors along the enemy line" (26). Wilson speaks truthfully here. His science is precisely what one would expect of a man who thinks this way. "Whiff of grape-shot" is his phrase, not mine.

The third southern value that I want to highlight is that of our old friend progress. Again, as Wilson himself repeatedly stresses, this was very much the philosophy of his youth. Thanks to the New Deal, Alabama and other southern states were being hauled up from the most appalling poverty. There was a reason why southerners voted for Franklin Roosevelt: massive state projects like the Tennessee Valley Authority were transforming people's lives and filling them with hope for the first time since the Civil War. It was natural for a young man to think that change for the better is possible—is actual—and can continue with good will and effort. This vision floods everything that Wilson writes. Nor should one think that his interest in the balance of nature is in any way antithetical. We have seen how from Spencer—a Wilson hero—on, people have combined a belief in upward progress with a natural tendency to balance: dynamic equilibrium.

Consider Wilson on the matter of biodiversity (1992, 187):

> Biological diversity embraces a vast number of conditions that range from the simple to the complex, with the simple appearing

first in evolution and the more complex later. Many reversals have occurred along the way, but the overall average across the history of life has moved from the simple and few to the more complex and numerous. During the past billion years, animals as a whole evolved upward in body size, feeding and defensive techniques, brain and behavioral complexity, social organization, and precision of environmental control—in each case farther from the nonliving state than their simpler antecedents did. More precisely, the overall averages of these traits and their upper extremes went up. Progress, then, is a property of the evolution of life as a whole by almost any conceivable intuitive standard, including the acquisition of goals and intentions in the behavior of animals.

Meaning through progress is a Wilsonian creed and it shapes his science. *Sociobiology: The New Synthesis* is as progressionist as Julian Huxley's *Evolution: The Modern Synthesis*. It gives us a progress from the colonial invertebrates, through the insects, fish, reptiles, mammals, apes, and finally to "man." The autocatalytic model of human evolution may (or may not) have the virtue of predictive fertility. It is certainly a crucial aid in Wilson's cultural commitment to an upwardly directed reading of life's history.

I have mentioned three aspects of southern life which influenced Wilson's science. But I am sure some critics will complain that I have missed the most important of all—its pervasive and inherent racism during Wilson's youth. Here, they will say, we might surely expect to find nonepistemic forces—values—influencing Wilson's work, along with sexism and classism and much else besides. Nor, in the opinion of critics like Lewontin, do our expectations fail. They claim that Wilson's writings on humankind endorse all of the biases of his group—white, middle-class, heterosexual males from the South. In *Sociobiology*, for instance, there are claims about males being dominant and aggressive and females being passive and receptive, about the biological naturalness of the nuclear family, about intelligence being under the control of the genes. Wilson never comes out and says that blacks are biologically inferior or anything like that. But to his critics, the ground is prepared, and it is dug again and fertilized in later books. This quotation exemplifies their concerns: "The evidence is strong that almost all differences between human societies are based on learning and social conditioning rather than on heredity. And yet perhaps not quite all." When Wilson then tells us that Chinese-American newborns are more placid than Caucasian-American infants, critics feel that we have the thin end of a very large wedge that others have been happy to hammer in much more firmly (Wilson 1978, 48–49).

This is all such an emotive and much-discussed issue that I am not sure one can say anything to bring things to a successful resolution. What I can say with certainty is that rank charges of racism or sexism are unfair and misplaced. Indeed, if anything, I give Wilson credit for having moved so far from childhood influences and his undergraduate years at the segregated University of Alabama: there, he, like all the other white heterosexual men he knew, would have assumed that blacks are coons, gays are pansies, and women are girls. What one can also say, nevertheless, is that the Wilson of *Sociobiology* jumped right in without much thought and, at the very least, was prepared to entertain judgments which would have been anathema to Gould and Lewontin. Since their positions and scientific styles were culturally influenced, it seems uneven to deny that Wilson's position and scientific style are likewise culturally influenced.

Indeed, more than just being subconsciously influenced by militarism, fundamentalism, and what some would regard as racism/sexism/classism, in many respects Wilson feels comfortable around them, talking about them. This was the case in the mid-1970s when he started to write about human sociobiology. It is not, for instance, that he thinks homosexuals bad; but he does think that they are different—different enough to require a biological explanation. Moreover, the implication is that, seen from an evolutionary perspective, homosexuals are handicapped or playing catch-up, since their existence is explained as a function of balanced heterozygote fitness (with homosexuals losing out to their super-heterosexual siblings) or of kin selection (with homosexuals helping their heterosexual siblings to reproduce) (Wilson 1978).

Wilson's Science in His Own Time

With a man like Wilson, whose cultural values and beliefs are so strong and so explicit, one wonders very much about the status of his science—both as he would have it judged and as others would judge it. Much of the work is professional by any reasonable standard. The work on biogeography, for instance: the data gathering and theorizing from the travels in Melanesia; the joint theoretical achievements with Mac-Arthur; and the experimental studies with Simberloff. Likewise, the later work on the ants: caste structure, division of labor, and the like. Nonepistemic values may not be entirely absent—one certainly senses approval of the division of labor—but the work is driven by the desire to be predictive and so forth. Something like adaptation certainly has its roots in culture—Christian culture—but by the time Wilson uses it, it has lost much of its emotive force.

However, as one turns to sociobiology, the nonepistemic factors start to rise, and conversely the epistemic constraints sometimes get looser. This is certainly the case when Wilson treats of our own species. Wilson speculates, for instance, that religion is a crucial part of the human psyche. "In the midst of the chaotic and potentially disorienting experiences each person undergoes daily, religion classifies him, provides him with unquestioned membership in a group claiming great powers, and by this means gives him a driving purpose in life compatible with his self-interest." Hence, "The mind is predisposed—one can speculate that learning rules are physiologically programmed—to participate in a few processes of sacralization which in combination generate the institutions of organized religion" (Wilson 1978, 188). At points like this, nonepistemic urges are substituting for evidence conforming to strict epistemic norms.

Wilson himself realizes this. Although *On Human Nature* is included in a trilogy starting with the very professional *Insect Societies*, Wilson carefully qualifies its status at the beginning of the book. We learn that it "is not a work of science; it is a work about science, and about how far the natural sciences can penetrate into human behavior before they will be transformed into something new. It examines the reciprocal impact that a truly evolutionary explanation of human behavior must have on the social sciences and humanities." Rather than professional science, we have a "speculative essay" (x). And, this is essentially how the book was received. It went quickly into a cheap paperback edition and (after receiving the Pulitzer) became a bestseller.

One doubts that Wilson himself particularly cherished the non-professional status of *On Human Nature*. Rather, it was that (under criticism) he realized that he could not hope to pass the book as one for the professional. With later works, the popular status seems more intentional and less grudgingly accepted. *The Diversity of Life* is written expressly for the general public, with flowing prose, no heavy material (like mathematics), and lots of glossy color photographs. Easy on the epistemic and strong on the cultural. Its commercial fate tells all. Trade publishers simply do not shell out half a million dollars for the paperback rights of books in professional science. Wilson's most recent books, the autobiographical *Naturalist* (1994) and *Consilience: The Unity of Knowledge* (1998), are also lucrative works designed for popular consumption.

Two Temperaments, Two Philosophies

Richard Lewontin and Edward O. Wilson have been ambitious and successful leaders in their field. In their work we have seen a mixture of the epistemic and the nonepistemic or cultural. However, there is now a strong tendency to confine the too-overtly cultural to the realm of popular science; often (usually) in the professional work, cultural values operate at the meta-level, where they do not so much enter into the science as affect the science one finds epistemically acceptable.

Lewontin's Jewish identity and his feeling about Nazi atrocities lead him to erect stiff standards for science, especially science about humans. Wilson's admiration for military bravery leads him to admire dashing and daring scientific hypotheses. Combined with his urge to find an evolutionarily based secular religion, this makes him more receptive than Lewontin to a biology of human nature.

Conclusions like this make one itch to return to questions of objectivity and subjectivity; but first let us complete our survey of evolutionists by turning in the next two chapters to men in their prime.

10

✍

G EOFFREY PARKER

The Professional's Professional

Spending your days out in the cow field, waiting for the brutes to defecate, is not most people's idea of a perfect camping holiday. But it was precisely this which led to one of the most celebrated pieces of evolutionary research in recent years. We are a long way from the popular perception of evolutionists at work—Indiana Jones types out in the desert digging up dinosaurs—but I hope to show that even cow pats and their humble denizens have their attractions.

Geoffrey Parker (b. 1944) was a student, both undergraduate and graduate, at Bristol University, in the West of England. On graduation in the early 1970s, he got a job at Liverpool University, where the head of the department was Arthur Cain, deservedly well-known for seminal studies on the adaptive nature of snail-shell bandings (Cain and Sheppard 1950, 1952, 1954). Parker is still at Liverpool, now running his own group on evolutionary biology, respected both for his close readings of nature and for the ease with which he moves into mathematical modeling, trying to make sense of what he and others have discovered about the world of organisms. Although he lies outside the Oxbridge circle, he was the first of his age cohort to be made a fellow of the Royal Society, an honor which no one questioned.

Dung Flies

Parker first made his mark through a series of papers stemming from his thesis project on the behavior of one of nature's less prepossessing members, the dung fly, *Scatophaga stercoraria* (Parker 1969, 1970a–g, 1974a). In this species, males seek out fresh cow pats, females then arrive and are mated by males, the females lay fertilized eggs on the pats, and the hatched larvae bury down into the feces, drawing their nutriment from the rich environment in which they find themselves. On this basic pattern are imposed all sorts of variations and adornments, and it was in these that Parker found much scope for his scientific labors.

His main focus was on sexual selection stemming from male competition, for the males outnumber the females four or five to one. This means that the males have to be diligent in their search for mates. They cannot simply leave things to chance but must place themselves in the best available position around or on the pat in order to have the best possible chance of finding a female. The best position is a function of the environment (very fresh pats tend to be too liquid for the females' safety), the positions of the females, and of course the density and positions of fellow males. "Males should be distributed between zones in such a way that all individuals experience equal expectations of gain. Hence the proportion of females captured in a given zone should equal the proportion of males searching there, assuming that all females arriving are equally valuable irrespective of where they are caught" (Parker 1978b, 219–220). This prediction is found to hold: the difference between the expected and observed figures is not significant (see figure).

As time goes by, the males' best strategy changes. Those who have found mates, having copulated (this takes half an hour or more), will start steering their females to the pats (if they are not yet on them), where the females will lay (oviposit) their now-fertile eggs, the males still attached over them in a kind of guarding stance. Those who have not found mates, or who have finished with a previous mate, will be still searching, either for a free female or for a copulating (or post-copulating) pair, where a struggle for possession may now take place. Theoretically, what one expects is that, with time, searching males should change their search area away from the wider surroundings of

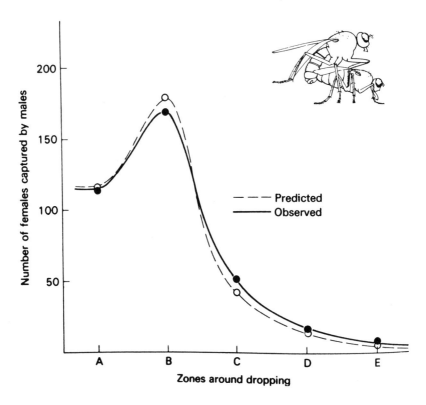

Comparison of observed and predicted number of female captures by male Scatophaga in each of a series of zones on and around cattle droppings (from Parker 1978b).

a field and toward the pats themselves, where a skin will have formed, making the pat less of a physical hazard for the females, who will by now be ovipositing on the pat.

Again there is a fit between observation and theory (see figure opposite). However, the fit is clearly not quite as exact as before. One expects a more rapid move toward the pat than one finds in practice. Parker suggests that this may be a function of the difficulty of getting information about the new droppings—until a skin forms on the earlier droppings, their smell will crowd out any new smells. But if the insect lingers about upwind, then it is perfectly positioned to detect new droppings, and the more accurate information compensates for the time used in obtaining it (225).

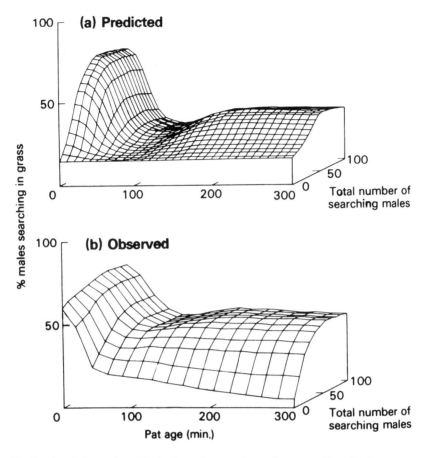

Predicted and observed profiles for Scatophaga male search strategy (from Parker 1978b).

Why is it in the reproductive interests of a male to attempt to remove an already-mating male from a female? Here Parker turned to experiment, working in the laboratory. He sterilized selected males and then allowed them to mate with a female; next, males that have not been sterilized were allowed to attempt to remove the first male and take over the female. By counting the viable offspring, Parker could calculate the proportion of eggs fertilized by the last (as opposed to earlier) mates of the female prior to oviposition. The proportion was a remarkable 80 percent (1970b). In other words, it really pays a male to take over a female, if he can, and it pays a male to prevent such a takeover, if he can.

Parker speculates that this latter fact is a major reason why, after copulation, the male stays on the female, guarding her (as it were) until the eggs are laid. Mechanical methods of protecting one's sperm, such as the mating plugs found in other species of insects, are not feasible (for morphological and physiological reasons) in dung flies. It is simplest, and the best Darwinian strategy, for the males to stay around until there is no threat from competitors (Parker 1970f, 785).

One other question of interest should be raised. If males so outnumber females, would it not be in a male's interests, once he has found and started to mate with a female, to keep going until he has fertilized all of her eggs? Surely, the more sperm in the female, the better? But balancing this is the fact that, while copulating with one female, he cannot be copulating with another, and it is the last copulator who really hits the reproductive jackpot. In other words, at some point time is better spent searching for new mates than sticking with old ones. In fact, one can set up a simple model, balancing search time against cost of time spent in copulation, showing the optimal strategy. The prediction is close to the observed average time of copulation (see figure). Parker suggests that the slight discrepancy may be due to the fact that the model does not take into account the cost to the male of producing sperm: this would presumably require time spent searching for food (Parker and Stuart 1976; Parker 1978b, 231).

Evolutionarily Stable Strategies

Parker's early thesis work was done essentially in isolation from other people who were starting to tackle the same sorts of problems as he. It was only after the bulk of his research was completed that he encountered John Maynard Smith's thoughts about evolutionarily stable strategies (ESS)—ideas which he embraced with enthusiasm, reinterpreting some of the dung fly results in their terms. At the same time, he started to move on to new issues, particularly those which lend themselves directly to ESS treatment, notably, animal aggression and the reasons why sometimes battles escalate and why at other times they do not.

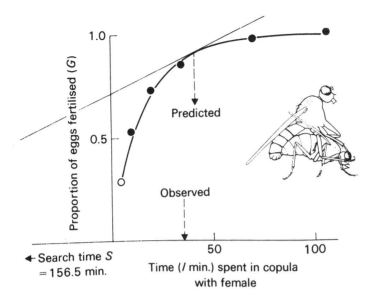

Optimal copula duration in Scatophaga (from Parker 1978b).

"Selection for aggression will be more intense the more discrete the resource (i.e. the easier it is to guard) and the higher its yield as a fitness gain parameter (a function both of its absolute effect and its shortness of supply). It is not surprising therefore that most of animal aggression relates to food fighting and especially to mating" (Parker 1974b, 224).

Typical of the kind of questions that Parker asked were those about the imbalances one might expect to find between organisms already in possession of resources of value (like territory) and those without. Which would be the more likely to escalate a confrontation and which more likely to withdraw, if not gracefully, then at least without too much cost expended on what might be a fruitless contest? The predictions are that when organisms are fairly evenly matched, combat should escalate, and that where there is imbalance in rank the higher or more powerful organisms are going to be more likely to push things up the confrontational scale and the lower or less powerful more prepared to withdraw relatively quickly. The predictions are strongly confirmed, as is the expectation that usually size will be a major factor, with the bigger outranking the smaller. One consequence of this last point is that usually males outrank females, although females with young tend to go up in rank. Expectedly, experience pays, as also does prior possession of resources (Parker 1974b, 240–241).

Resource competition was a relatively short-term interest, for soon Parker was immersed in another enthusiasm: parent-offspring conflict. Referring to work by the American sociobiologist Robert Trivers, Parker agreed that the reproductive interests of parents and offspring could well differ, with parents selected to give less to any one individual (in the interests of the whole litter or brood) and offspring selected to strive for the maximum, even at the expense of siblings. There would as a consequence be a parent-offspring conflict over the amount the parents would give or invest in any one offspring (Parker and MacNair 1978, 97).

As it happens, this whole idea has been bathed in controversy right from its beginning, with some biologists arguing that the very notion of conflict is virtually contradictory (parents are bound to win, no matter what) and with other biologists arguing that the notion is extremely fruitful (if anything, the scales are tipped in the direction of children). Parker's answer was that "it all depends." Distinguishing between conflicts where parents are faced with different broods and help given to one diminishes help given to another (interbrood conflict), and where parents are faced with conflict within a brood for a fixed amount of resource (intrabrood conflict), it turns out that the kids usually get the upper-hand in the former case and the parents in the latter. Trying to balance two or more broods is beyond parents, but they stay on top when faced with just one batch of hungry mouths (Parker and MacNair 1979, 1211). (Shades of human life here?)

When both parents contribute to child care, conflict arises between the parents as well as between parents and children. Each parent is trying to get the other to give more to the present offspring, thus freeing up its own contribution to be devoted to other offspring. And the result is that the offspring get less than if only one parent were contributing. "How does this paradox arise? The reason is straightforward: if two parents collaborate, the effects of a unilateral reduction in investment by one parent would be partly offset by the other parent: the costs are felt in full by the single parent" (Parker 1985, 522).

Parent-offspring conflict led naturally to discussions of clutch size, with much interest in siblicide, where, as in golden eagles, one chick frequently or always kills nest mates (Godfray and Parker 1991, 1992). Once again we have matters centering on relative reproductive strategies—parents who want as many viable offspring as possible and siblings who want as much as possible for themselves, compatible with not doing themselves a genetic injury by killing off fellow gene bearers to no good purpose. Work on this sort of problem has continued right through to the present, and clutch size has been an ongoing concern.

I want to conclude my exposition of Parker's science with a brief glance at a paper recently coauthored with the Cambridge sociobiologist Tim Clutton-Brock, a man who established his credentials as an important evolutionist with a study of red deer on an island off the coast of Scotland (Clutton-Brock, Guinness, and Albon 1982).

Punishment

The topic is punishment, "the infliction of an ill suffered for an ill done" (Clutton-Brock and Parker 1995a, 209, quoting Grotius). As with much of Parker's work, this is not a new notion for the student of animal behavior, but in his hands it gets refurbished and is seen as a more stimulating and productive area of study than was hitherto realized. "Individuals (or groups) commonly respond to actions likely to lower their fitness with behaviour that reduces the fitness of the instigator and discourages or prevents him or her from repeating the initial action" (209). It can be shown how punishment is but one of a number of interactions possible among nonkin and how one would expect fitness benefits to flow. Most particularly, one can show that under certain plausible assumptions, punishment strategies will achieve equilibrium: they will be ESSs.

But when and where does one expect to find punishment as a regular practice? In many places and often, apparently. Punishment occurs most obviously in dominance relationships, with the dominant tending to punish their subordinates to keep them in line, although sometimes if the subordinate has nothing to lose, the tables are turned, with the subordinate attacking the dominant animal. One also finds punishment occurring in sexual encounters. Apparently here it is all very much a question of keeping the females in line. Jane Goodall has described how male chimpanzees beat up females to make them more compliant and to teach them not to stray too far. Thus the male keeps his potential mating partners under his control.

In some species, apparently, the male uses punishment when he is not granted expected sexual favors: the male uses violence toward those it thinks should be or would like to have as partners. "In polygynous or promiscuous species, males sometimes attack females that refuse to associate with them. Territorial male red and fallow deer will prod straying females with their antlers. Male hamadryas baboons initially threaten females that stray with an eyebrow flash but if they fail to return immediately, will bite them on the neck" (213).

As one might expect, the authors show interest in the role played by punishment in parent-offspring conflict. And as the discussion draws to an end, they speculate more and more on punishment in our own species (215):

> It is clear that analogies exist between punishing tactics in social animals and retaliatory aggression among humans. Retaliation is a common feature of aggressive interactions between individuals as well as between groups: obvious examples include retaliatory expulsions of diplomats and tit for tat sectarian killings. In some tribal societies, individuals are permitted to punish minor infringements of their interests themselves, while the punishment of more severe offences is either carried out by community action or is delegated to particular individuals, an arrangement likely to limit the extent of negative reciprocity and minimize disruption within the group.

Analogies are drawn with aggression in the hymenoptera as well as in the primates, and on this note we end our exposition of Parker's work.

Values in Parker's Science

Parker is an unabashed Darwinian adaptationist. Whenever and wherever possible, he looks for function as molded by natural selection. Very revealing in this respect is a little paper he once wrote on the mounting behavior sometimes shown by female ungulates. Someone like Gould would have assumed at once that this was nonadaptive, simply a by-product of the fact that males and females are built on the same plan: androgens and estrogens occur naturally in both sexes, and if a bit of a slop-over occurs one way or the other, this is only to be expected.

Not so for Parker, whose inclination is to find the behavior linked to natural selection. He floats the suggestion that the time especially when females mimic male sexual behavior is when they are in heat, and the reason for both the behavior and the timing is precisely because this is a very powerful method of attracting the attention of dominant males. Seeing someone else on the job—one's own private personal job!—is a sure-fire way of "stimulating its aggressive responses" and getting it to come over and discover that the mimic herself is ready for mating (Parker and Pearson 1976, 241–242).

I give this example not because Parker's interpretation is necessarily right but because it shows how very much Parker works within the adaptationist framework. He is not insensitive to this fact—how could he be, given the criticism of adaptationism in recent years?—and has referred to it explicitly, defending his way of thinking about evolution. In particular, he makes the crucial assumption that selection maximizes or in some sense optimizes the adaptations that one finds in nature. About this, in an article coauthored by John Maynard Smith, Parker has admitted candidly that any question they ask "is assumed to have an adaptive answer, otherwise we cannot proceed to establish whether a given adaptive process can generate the correct solution" (Parker and Maynard Smith 1990, 27).

But this does not mean that nothing further needs to be done by the evolutionist. First: "A range of alternative actions or 'strategies' relating to the question is defined" (27). Then come the nuts and bolts of model building. We must decide what is being maximized or optimized. "The simplest direct criterion is the expected lifetime number of surviving offspring produced by an individual pursuing a given strategy ('individual fitness') defined in units of generation time (which may vary with phenotype)" (27). With this come the assumptions: "Once the optimization criterion has been chosen, assumptions have to be made about the fitness consequences (or 'payoffs'—a term borrowed from game theory) of the different strategies, which may involve construction of mathematical models" (27). And so, finally, we are in a position to make inferences: "Once payoffs to the strategies have been stated, the optimal solution(s) are deduced by an appropriate analytical technique" (29).

Parker is acutely aware that this is not the only way to think about evolution, even evolution under the strong control of selection. Two other methods offer possibilities: the comparative method and the method of quantitative genetics. The former, which should be thought of as a complement rather than a rival to optimization, goes back to Darwin. Here one analyzes and compares features across a range of organisms. This is just what Clutton-Brock and his associates did in deciding on the causes of sexual dimorphism in primates. They preferred the answer based on selection arising from competition between males for mates (the biggest got the females) to the answer based on a division of labor (the males are bigger because of the special tasks that they perform). The reason for this choice was that, in monogamous species of primates, the dimorphism is much reduced. Monogamy minimizes male-male competition, and even though the tasks and roles of the two sexes remain different, males and females in monogamous relationships are much more evenly balanced in size (Maynard Smith and Parker 1990, 32).

The other possible approach (the Lewontin-type approach) is genetic. About this, Parker writes: "Genetic models have the advantage of greater realism and also allow the study of evolutionary dynamics as well as evolutionary end points. However, genetic models entail the explicit assumption of genetic mechanisms, almost always in the absence of any real knowledge of the underlying genetic basis of the trait" (Godfray and Parker 1992, 484). Even if he himself does not often work at the genetic level, Parker is much aware that he is making genetic assumptions. The extent to which natural selection will be able to optimize its actions will be a function of the effects that the genes have on their carriers and of the facts of genetic transmission, especially at the populational level (Maynard Smith and Parker 1990, 30).

Recently, as opposed to his early years, Parker has been working more at theory and model building than at observation in nature and experimentation in the laboratory. But one gets the continued impression—the very strong, continued impression—that one is dealing with a man of scientific integrity, of epistemic purity indeed. In Parker's world, one cannot simply make assumptions and inferences in a loose way, thinking that they will go through and hold. Mathematics does count, and one has an obligation to make one's inferences and model building as tight and formal as possible. Norms like consistency and coherence are paramount.

By his own admission, Parker is not given to grand system building, preferring rather to work away on specific problems. Hence, we should not look for (nor will we find) overriding consiliences. Parker is sensitive to the need for unity and to the way in which his work fits into the overall Darwinian picture; but more than this is not his job. Fertility, in the sense of setting out self-consciously to provide a whole new direction of research, also is something he rather takes for granted and is happy to offer as the case may be; but he is not another Wilson. The article on punishment is certainly a call to inquiry, but it is set severely within already-articulated theory. In his early work, Parker pioneered his own individual selectionistic way, but probably the most important concept-structuring work he has done is that of an evolutionarily stable strategy—an idea worked out by others, notably Maynard Smith. Added to this are the ideas of Robert Trivers, especially parental investment and the consequences for parent-offspring conflict.

My point is that, to use Kuhnian language, most of the time Parker has worked within the overall Darwinian paradigm; he has been a normal scientist doing normal science. Solving puzzles rather than tackling problems. A pragmatic, one-step-at-a-time kind of scientist. Parker tends to be wary of ideas that attract attention purely or mainly on the basis of their beauty or elegance. No scientist can be indifferent to this value, but Parker is much aware of the temptation to find the evidence to fit the theory that one is attracted to on aesthetic grounds (Ruse 1996).

I am describing Parker as a meat-and-potatoes kind of scientist, a description I suspect he would accept happily, with pride even. But if the meat is theory, then the potatoes must be the norm of predictive excellence. Do the inferences fit the facts? And it is here, certainly, that

Parker would find the heart of good science. Does it work? Does it yield true predictions? Listen to Parker writing with Maynard Smith (Maynard Smith and Parker 1990, 29):

> The final step in the optimality approach is to test the predictions, quantitatively or qualitatively, against the observations. If they fit, then the model may really reflect the forces that have moulded the adaptation. If they do not, we may have misidentified the strategy set, or the optimization criterion, or the payoffs; or the phenomenon we have chosen may not in fact any longer be adaptive. By reworking our assumptions, we modify our model and revise and retest the predictions.

From the beginning Parker's aim has been to make empirical predictions that correspond to reality. This was the driving force of the work on dung flies. What does the theory tell us about something like the optimal mating time for male flies? What do we find in nature? Are the two answers close? Are the answers close enough that we can conclude that the theory really is telling us what is going on in nature? If the answers are significantly different, have other factors been ignored that could possibly account for the discrepancies? And so forth.

The strong replies to these questions have given Parker good reason to feel that his is a successful predictive science. Undoubtedly, behind the scenes of the dung fly work lies a fair amount of effort, adjusting theory to match the findings. As Parker and Maynard Smith admit, this is the way one does science. "By reworking our assumptions, we modify our model and revise and retest the predictions" (29). But the result has been science which scores well on that crucial epistemic value of prediction—the value which is so prized in the physical sciences and so often missing or unfulfilled in earlier work in evolutionary biology. When I wrote of Parker as doing normal science, I meant this as praise, not as criticism.

Yet Parker is not a scientist of the public domain. The son of a professional chemist and the younger brother of a physicist, Parker publishes his work in professional journals intended to be read exclusively by professionals. He has written a few review articles in *Nature*, but these tend to be at the technical end of the scale. By his own rueful admission, his only attempt at a popular article (for *New Scientist*) had to be virtually ghost written (Baker and Parker 1979). He has resisted subsequent entreaties that he try his hand at writing for the general public.

One should therefore not expect to find—and one does not in fact find—broad-scale nonepistemic values intruding in Parker's work, of the kind one finds in Wilson's *On Human Nature* or Lewontin's *Not in Our Genes*. Of a friend who has written a popular book on sperm competition, a topic which Parker rightfully thinks he himself pioneered, Parker reflects almost primly that although he does not "blame" the friend for writing it, "it's almost pornography in its way." Ambivalently, Parker adds: "If he can earn a fast buck out of it, jolly good luck to him" (interview with author, May 1997; all unattributed quotations from Parker that follow are from this interview).

So is Parker's work finally an example of the "pure science" that the objectivists praise? Is it beyond culture? Here I would recall the distinction between full-blooded cultural values and factors or elements which are not less cultural but not thereby necessarily value-impregnated. Throughout everything he produces, Parker relies heavily on the models and metaphors and examples of his predecessors, brought up to date by the culture of his day. The Darwinian picture is rife with cultural elements—selection, adaptation, function, and much more. Add to this the very heavy use of game theory—developed not just by gamblers but also in recent years by the military and business and other major players in our society—and you have theorizing that is as indebted to the late twentieth century as any novel that wins the Booker or Pulitzer Prize, or any Oscar-winning high-tech movie or gleaming all-glass corporate headquarters. Evolutionarily stable strategies are to biological science what computers are to airlines—improvements.

In this sense, Parker's work is deeply cultural. It is cultural also in style, or at least it was for the early dung fly papers:

> I found it difficult to write papers originally because I'd read a lot of William Faulkner's novels. I used to feel that the punch line in the paper had to be worked towards. In other words, you didn't have to give a clue as to what the answer really would be until you

got it. And this was a real problem actually in science. I don't do it now. I've tried to copy more the Maynard Smith style where you lay things out very logically and say what you're trying to do and what you feel the answer is, and then show how you get it. But in those days, I felt you had to build to it.

This is all interesting, but is any of this truly suggestive that Parker's science is (or has been) cultural in a way that would upset objectivists and delight subjectivists? What about values? Is there any evidence at all that Parker is sliding in wishes and preferences and prejudices by the back door, while pretending to be giving an account of objective reality? The answer is certainly that he is not doing this in any obvious way; the mere use or influence of cultural elements does not imply approval or values. It is hardly plausible to suggest that the steamy sexuality of Faulkner's Yoknapatawpha County is being read in or out of the cow patch.

In fact, one can go further than this. There are positive reasons to think that cultural values do not figure in Parker's work. He admits to little enthusiasm for that old favorite, progress. As one who read and was influenced by Kuhn, he is not even sure that he subscribes strongly to scientific progress. But neither is Parker much of a booster of the other side, that of God's working His purpose out: Providence. Take the question of adaptation. Parker certainly came by it honestly. As it happens, *pace* Lewontin, his mother's early experiences of working in a money lender's office had set up a family fear of borrowing (including from mortgage companies), which decreed that Parker's early life was more humble than his father's status and income might have decreed. But parental enthusiasm for natural history was reinforced by the sincere church-going Anglicanism of the mother and the passionate Methodism of the family of his closest friend. For Parker, the living world in all its Anglo-Saxon adaptive glory is part of his very being: "I do feel very English in a way. Rolling meadows, Friesian cows, oak trees. It's very much a thing I'd find hard to do without."

Yet, far from Parker seeing his science as surreptitiously supporting and approving of Christian natural theology, he argues that his science is a way of clearing out the fears and prejudices and hatreds and superstitions brought on by religion. Although he admits to a sentimental attraction to the church of his childhood, in his teens religious belief started to fade and has never revived. Speaking of the recent death of his wife—"the worst time of my life"—he feels that paradoxically the lack of religious belief made things "slightly easier" rather than other-wise:

> I didn't feel I had to ask the question "why" in a philosophical sense. I think if I'd been religious I'd have been tormented by the question of why had God done this to me. But that was never a thing. I just accepted that there'd been some transformation in some cell originally that had caused a cancer, and then nine years after the operation in 1985 there'd been a metastasis which had brought about her death. You can ask why do cancers form, but that's not a question you can feel a bitterness towards.

Parker stresses that he can be moved to deep emotion by music and poetry. Does his attitude mean that, as for Wilson, science is a kind of religion substitute? "Science is not a religion but it is slightly more than just what it would purport to be . . . It's a liberation really. You can choose to be highly superstitious, to believe in all sorts of things if you wish to. Some people erect huge superstructures of superstition—they can't allow themselves to go out on Friday the thirteenth. They must do this before they go to bed. Whatever. Salt over the shoulder. Mustn't wear green. And it goes on and on. I think really in a way I do see science—being a scientist perhaps—as some sort of liberation from those things."

Of course, in a sense, one might argue that this shows that for Parker (as for other scientists) his science has the metavalue of being prized precisely because it is objective and not (culturally) value-drenched. But he and other objectivists would surely accept this, for it would not be seen as compromising the integrity of science. The very opposite in fact. Parker himself, in discussing human sociobiology, draws a distinction between objective science and subjective desires. One has to take the genes into account when one is deciding moral issues, but it would be wrong to think that the biology itself dictates the moral course of action (Parker 1978a, 854):

For instance, if men are shown to have genetically-based shorter lives than women, it seems appropriate to consider this in a discussion of retirement age. Failure to heed genetic fact can here cause unfairness. But suppose it were proven that I am genetically predisposed to bequeath all possessions to my son at my daughter's expense. In this case, unless it were unequivocally established that an egalitarian distribution causes asymmetric hardship, I would not wish to heed genetic fact. Genetic predisposition itself seems irrelevant to "right" decisions; the important question is whether ignoring a predisposition can cause injustice.

Choosing Topics

Let us press our inquiry. What about choice of topics? Does this show values in some sense? Anyone who has searched for a PhD thesis topic knows full well that all sorts of factors enter into the choice, not the least of which is whether the topic represents an open ecological niche: ideally one wants a topic on which not too much has been written, or at least not written well. Sexual selection certainly fit that bill in the late 1960s. But as a matter of fact there was more to Parker's choice than this. He is candid about the extent to which personal factors have influenced the choice of topics on which he has worked. "Let's face it. Every human male of twenty, twenty-five, is mainly preoccupied with sex . . . So I don't think it was entirely random that the first thing that I wrote on, really began work on, was male-male competition and sexual selection. And I think possibly sexual conflict and parent-offspring conflict. Maybe I was more fascinated by doing parent-offspring conflict because I had young children . . . So it was part of personal life. Sexual conflict? Well! [laughs] I suppose all males have been in that position occasionally."

Although not trained formally in philosophy, Parker has a keen interest in the methodology of science, and so he continues self-reflectively, "I think one tries to be as objective as one can, but in the end in constructing a model you have to make the right assumptions. You've got to have some intuition into the way life is, to construct a model. And that is one of the most difficult steps . . . It's the formalizing, the conceptualizing, what's the nitty gritty of the whole thing. How shall we set it all up and how shall we get it in a sensible and coherent structure? That's quite a hard step. And I think that in that step you have to have some intuition into the way life is."

This is a crucial process which Parker likens to the creative act of painting a picture. It means that the scientist draws on life experiences and the lessons learnt from them:

A theorist, a model builder, has to have some concept of the way life is and above all he has to have some question. So you have to construct some model, some mathematical model, in such a way as to be intuitively satisfying, such that it can produce some solution to the question you're trying to answer. That's the hard step. That's what makes good models and bad models. Whether they're conceptualized right, whether they're framed right. In a sensible sort of way. And I don't think doing that is done *in vacuo*. You've got to use your intuitions, your common sense, your feel for biology. And I am sure that life as it has happened to you has some bearing on that. It can only have. I suppose parent-offspring conflict, perhaps sexual selection, sexual conflict, are all things which we have some intuition for.

But still, this philosophy of research does not necessarily mean that Parker's values intrude into his science. The work on parent-offspring conflict does not mean that he approves of it, or that he takes one side rather than the other. It could be rather that he wants to deal with it, seeing if it is harmful and, if so, whether it can be eliminated through an understanding of its true nature. Which point could certainly be true, although whether this means that values are absent completely is perhaps another matter. For all that Parker warns explicitly against moving from fact to values, from thinking that because something is natural it is therefore good or acceptable, one might feel that this is a criticism that could be leveled against his work.

Take the question of male-female differences and of their relative natures. The early Parker, discussing and analyzing his dung flies, puts maximum emphasis on the male role and activity, portraying the females very much as Victorian maidens awaiting their fates. Female activity—the rate at which they arrive and where they go, the willing-ness to be mounted, and so forth—is mentioned, but compared to the males, very little is said. The impression is that the males are the major players, even though evolutionary biology suggests that females ought to be working flat out to maximize their interests also.

Part of this distortion is artifactual: the males are the more visible actors, and, whatever the females are doing, it must be more subtle. Hence, there is good reason to work on males first. But the imbalance is not to be explained away so easily, especially since this imbalance persists in the recent work. In an article on harassment coauthored with Clutton-Brock, the males continue to get the bulk of the attention. And the female actions tend to be portrayed as responses to males (Clutton-Brock and Parker 1995b, 1360): "Male harassment may have profound consequences for the movements of receptive females. In a number of insects, females avoid areas where males are abundant in order to minimize harassment" (Parker 1970e, 1978b). Even when it comes to solutions, ways of avoiding harassment, the females have to accommodate to the males, staying in line and so forth. The very best solution seems to be to choose the right male—a dominant, powerful one—in the first place.

Even here the objectivist response might well be that this is an issue of social concern, so one can hardly fault Parker for being interested in it. He does not imply that harassment is a good thing, especially not in the human species. But the authors do speak of harassment as a form of punishment, and this raises some concerns. One would surely say that punishment in the human realm is morally right and obligatory—not always, but often. Who would want to say that Adolf Eichmann should have gone unpunished? However, if one is going now to extend the discussion to the phenomenon of males beating up females in order to ensure sexual compliance, one might well ask whether it is appropriate to continue to use such a morally laden term as "punishment." There is an all-too-easy chain of inference from "punishment in humans is good" to "punishment is natural" to "punishment includes beating up females for sexual ends," and the reverse inference back again.

Although this inference is not made explicitly by Parker and his coauthor, feminist critics have found it in his work. Expectedly, Parker finds this upsetting and to a certain extent hides behind Clutton-Brock.

"My interest really, I suppose, is more in the mathematics and the structure and function of the punishment model and of how it works and so on—and whether it really does relate [more] to primates than to humans." He stresses that both his parents and he himself have been rather mild and nonabusive people. "I had the occasional slap, but there were no beatings, nothing like that—it was a very gentle time. And I don't think I've been like that with the kids. I've never indulged in—I don't think I'd be physically suited to—battering other people at all."

The Scientist's Obligation

Parker spells out his position: balancing what he sees as the right, the positive, moral obligation of the scientist to follow ideas wherever they lead with the equal obligation to recognize the limitations of what science can achieve. "I think we must have the freedom to make models of natural systems that are plausible explanations. I think the punishment model is a plausible explanation of some behaviour we see in primates. What I wouldn't like to do is to commit what was always known as the 'naturalistic fallacy' and say 'What is, should be.' Because, quite frankly, if men end up beating up women, I'd have no sympathy for that at all. I'd be very glad to see them flung in jail for a very long period of time for doing that. This is absolutely no reason to justify such behaviour."

Parker's sincerity needs no defense. Yet I am not sure whether the passage just quoted sits entirely happily with what he says about the need of the researcher to have understanding of the premises of his models. And one might say, whatever his (or Clutton-Brock's) views, the very act of publishing such work in public places gives the ideas standing of their own. If those planning retirement policies can take account of genetic factors, why should not judges and social workers likewise note such factors? This does not justify family violence, but it suggests that there are moral depths lurking beneath even Parker's calm waters.

Probably, readers have already made their own judgments on the role of culture at this point, and there is little that I (or Parker) can say to change people's minds. What we can say is that, whatever the reservations, on the question of values in science, Geoffrey Parker is a long way from Erasmus Darwin. A very long way.

11

>≡<

JACK SEPKOSKI

Crunching the Fossils

In 1982 Ernst Mayr published *The Growth of Biological Thought,* a very long and massively researched history of evolutionary ideas. Being the man that he is, Mayr did not shrink from passing judgment on the biologists of the past. Some came in for great praise; others fared less well. But no one felt the scorn directed toward Herbert Spencer. Mayr could hardly bring himself to mention the man. "It would be quite justifiable to ignore Spencer in a history of biological ideas because his positive contributions were nil" (386). Perhaps so; but as the twentieth century draws to an end, Herbert Spencer is alive and well and living in Chicago.

J. John Sepkoski Jr. (b. 1948) was a graduate student at Harvard in the early 1970s, much under the spell of the lively young faculty member Stephen Jay Gould (Sepkoski 1994). Sepkoski's first post was at the University of Rochester, in the same department as the senior (G. G. Simpson-trained) paleontologist, David Raup. In the late 1970s both men moved West, soon again becoming colleagues at the University of Chicago, where Sepkoski remains.

A recipient in 1983 of the Schuchert Award of the Paleontological Society of America, given to the best young researcher of the day, Sepkoski has just completed a term as president of that body. And this in itself is worthy of note, for he is very much a new breed of paleontologist. I doubt that he has ever dug up a real fossil. Sepkoski is happiest when collecting vast libraries of data and then crunching the numbers through the computer, seeing what patterns emerge at the other end. A new age has arrived.

Fossil Biogeography

Sepkoski is no mere Baconian inductivist, hoping that some pattern—any pattern—will emerge. Like Parker, he works by model building, seeing if his ideas fit the facts, and adjusting and revising until he gets a reasonably close match between the evidence and the model. In order to do this, he needed some inspiration—some problem or obsession on which he could base his foray into the unknown—and so, for his first major project, Sepkoski turned to one of his Harvard mentors.

Not to Gould, though. "I'd been thoroughly schooled in the Mayrian school of speciation and so I read the manuscript [on punctuated equilibria] that Niles and Steve had written, and my reaction was that this is so obvious" (interview, June 21, 1997; in this chapter, all unattributed Sepkoski quotations are from this interview). The influence of Edward O. Wilson was different, however—his "was just aflterrific course, Wilson's a master lecturer"—and the MacArthur-Wilson (1967) theory of island biogeography really impressed. This was something that one could apply elsewhere, and indeed one of Sepkoski's first publications was precisely an analysis of the distribution of the freshwater mussels to be found in the rivers of the Atlantic coast (Sepkoski and Rex 1974). He and his coauthor took such rivers to be islands in a sea of land, and they tried to show how the MacArthur-Wilson theory gave one models of understanding.

Soon thereafter, Sepkoski was off into the past, following upon suggestions of Wilson himself (Sepkoski 1976, 298; MacArthur and Wilson 1967). If one thinks of the future (now a past-future to us) as a space to be colonized, what interplay would one expect, given certain specified rates of species innovation (corresponding to species arriving on islands) and of extinction (corresponding to species leaving or being wiped out from islands)? In particular, as in the MacArthur-Wilson theory, should one expect equilibrium? Simple equations of density-dependent rates of origination and extinction of taxonomic groups (taxa)—less rapid origination and more rapid extinction as pressures build up—show that, with respect to diversity, one expects overall to get a sigmoidal (S-shaped) rise in numbers of taxa and then a leveling off at a plateau of equilibrium (Sepkoski 1978, 233). After the initial growth, the equations suggest that the number of new species in any period more or less balances the number of species which go.

But is this a true picture of the fossil record? Unfortunately, species are not easy to trace in the fossil record. Usually one is forced to work with taxa of a higher level (see box). Looking then at the numbers of marine multicellular (metazoan) orders (a rather high-level taxon but the best that he could get at first), Sepkoski suggested that the record since the beginning of the Cambrian period, the Phanerozoic—570 million years in all (see figure page 218)—does indeed fit the expected pattern. Overall, we do truly have a sigmoidal curve and then rough equilibrium (234; see figure page 219).

For all its crudeness, the model yields a remarkable result, not the least in its implications for the happenings at the beginning of the period. The Cambrian explosion—an overflowing of life where just a short period before there was virtually nothing—has long been one of the great mysteries of evolution. Darwin, who knew nothing whatsoever of any pre-Cambrian life, admitted that it was one of the major lacunae of his theorizing. And now, although we are indeed aware of life before the Cambrian—life of a kind which an evolutionist would expect before and leading up to the Cambrian—why things should have taken off in such a spectacular way is still a major mystery.

Sepkoski's picture suggests that, spectacular though the rise may have been, it need be a mystery no longer. Once the metazoans had evolved, such a rise (followed then by balance) is precisely what one expects. Most people assume that for some unknown reason, the rate of taxon production must have accelerated dramatically as the boundary into the Cambrian was crossed. Such an assumption is not necessary (228):

> The fit of the exponential model [i.e. Sepkoski's application of the MacArthur-Wilson model to the fossil record] indicates that total rate of diversification certainly was accelerating across the Precambrian-Cambrian Boundary . . . However, this acceleration resulted

The Linnaean Hierarchy

Following the eighteenth-century Swedish botanist Carolus Linnaeus, taxonomists classify all organisms in a hierarchical system. There are seven standard levels or *categories*, although today there are many intermediate or sublevels which are often used for more subtle nuances of discrimination. Within categories, one finds the fundamental groups or sets of classification: *taxa* (singular, *taxon*). Every organism is assigned to one and only one taxon at each category level, and these taxa form nested sets, meaning that if the members of two or more taxa at any level are combined within one taxon at the next higher level, they must then stay together through the taxa at yet higher levels.

As an example of the Linnaean hierarchy in action, take the classification of the wolf:

Category	Taxon
Kingdom	Animalia
Phylum	Chordata
Class	Mammalia
Order	Carnivora
Family	Canidae
Genus	*Canis*
Species	*Canis lupus*

The taxon at the lowest standard level, a species, has a binomial name (which is italicized), starting with the name of the genus within which the species members fall.

There is no fixed number of taxa from one level which will be included in a taxon at the next higher level. We humans, *Homo sapiens*, are the only living representatives of the genus *Homo*, but before us there were the species *Homo habilis* and *Homo erectus*. The fruit fly genus *Drosophila* contains over a thousand species. The point of the hierarchy obviously is that more and more organisms will be in fewer and fewer taxa as one goes up the hierarchy. Hence, when Sepkoski moved from looking at taxa at the order level to taxa at the family level, his analysis was becoming more fine-grained.

	EON	ERA	PERIOD
			Quaternary
1.6 Mya		Cenozoic	Tertiary
65 Mya			Cretaceous
138 Mya		Mesozoic	Jurassic
205 Mya	Phanerozoic		Triassic
240 Mya			Permian
290 Mya			Carboniferous
380 Mya		Paleozoic	Devonian
410 Mya			Silurian
435 Mya			Ordovician
500 Mya			Cambrian
570 Mya			
	Proterozoic	Precambrian Time	
2.5 Bya			
	Archaean		
4.6 Bya			

The geologic time scale. The most recent analyses put the beginning of the Cambrian a little later, around 545 million years ago (Mya), than was thought previously. On this new dating, the period which is often marked out as preceding the Cambrian (and which is called the Vendian) gets brought forward accordingly.

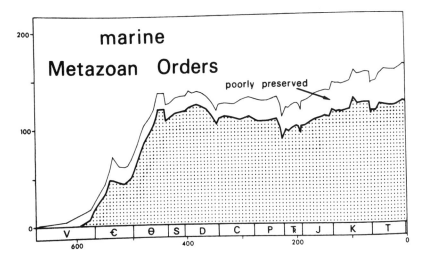

The complete Phanerozoic record of the diversity of marine metazoan orders (from Sepkoski 1978).

simply from the multiplicative effect of the continuous addition of taxa. *The per taxon rate of diversification seems to have remained constant.* Therefore, in the absence of change in this fundamental rate of diversification (i.e. r_4), there seems to be no reason to invoke any extrinsic trigger specifically at the Precambrian-Cambrian Boundary.

Sepkoski admitted that, in a way, he was not so much eliminating all of the questions posed by this part of the fossil record as moving them backward. We retreat in time from the Precambrian (Vendian)–Cambrian boundary toward "events and processes surrounding the initial appearance of metazoans in the fossil record near the beginning of the Vendian, more than 100 Myr [million years] earlier" (Sepkoski 1978, 228). We may no longer worry about the explosion in the Cambrian, but now we have the puzzling appearance of multicellular organisms in the first place.

Following others, Sepkoski suggests that this event may have been connected to the evolution of sexuality. However, this is not really the kind of question with which Sepkoski himself is very much concerned. He is hardly indifferent or hostile to causes, including causes of a Darwinian kind (that is, connected with adaptation), such as one might suppose for a new life form. But he is really more interested in the dynamics of large groups, however the members may appear and disappear. In this attitude we do surely see the influence of Gould, Sepkoski's teacher, who was just then arguing that new species appear rapidly, for fundamentally nonadaptive reasons, and that the overall patterns in the fossil record cannot and should not be tied tightly to the Darwinian processes of micro evolution. Another publication—which Sepkoski coauthored—argued that the shape of the evolution of groups (clades) is basically the same as if it had been generated randomly by a computer (Gould et al. 1977)!

Crude though the first model may have been, it was the foundation for further developments. In the years following, Sepkoski was able to expand his data base to the more fine-grained level of families, rather than orders. This showed at once that something was wrong with his previous, rather smooth, sigmoidal upward curve in the Cambrian. There is a kind of miniplateau in the middle of the Cambrian as the rise in diversity pauses, before it again picks up the pace. More than this, those organisms which really flourish in the Cambrian reach their peak at the time of this miniplateau or point of equilibrium, and then they go into a long slow decline (see figure; Sepkoski 1979, 235).

One needs, perhaps, two sets of equations yielding two curves, which would allow one to superimpose the second or later Paleozoic fauna on top of the first Cambrian fauna (Sepkoski 1979, 238). And as it happens, with careful manipulation of the figures, Sepkoski was able to get the desired pattern which fits the actual path plotted by nature. "The two-phase kinetic model . . . seems to provide an adequate description of the fundamental patterns observed in the early Phanerozoic diversification of marine metazoan families" (242). Note that this is still in essence a descriptive model, saying nothing of underlying causes. "The title I gave to the model—the 'kinetic' model—that word was very carefully chosen. This wasn't really about evolutionary arguments but rather a model describing a pattern without necessarily having all the underlying dynamics built into it."

However, the temptation to hypothesize about underlying factors proved too much. What we may be looking at is the different evolutionary patterns exhibited by "generalized" and "specialized" taxa. The for-

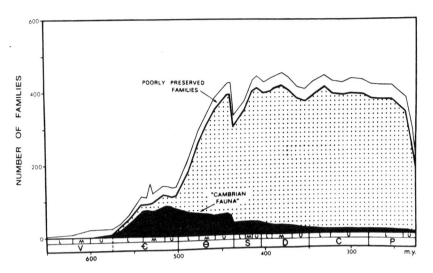

The complete stage-by-stage history of familial diversity through the whole of the Paleo-zoic, showing "multiple equilibria" with two intervals of logistic diversification. The black field represents familial diversity within those classes that are important in or are restricted to the Cambrian. The stippled field represents the diversity of remaining skeletonized families (from Sepkoski 1979).

mer, meaning taxa whose members have "relatively broad feeding and habitat adaptations," might be very good at rapid diversification into essentially empty ecological niches. But they run out of steam as the space gets filled up. The latter, specialized taxa, "might be expected to exhibit lower rates of speciation and extinction and, as a result, lower rates of diversification but higher equilibria" (Sepkoski 1979, 243). They concentrate their energies on specialized niches and opportunities. What they lose on the general front they gain when the going gets tough. Thanks to their distinctive characteristics, the end result might be "more finely divided and stable ecosystems which can be described as having high equilibrial diversities" (243).

This is starting to sound a little bit like good old-fashioned pro-gressionism. Sepkoski seemed to be aware of this fact and offered a careful qualification (244):

This attempt to characterize the average ecologies of the two early Phanerozoic faunas as generalized or specialized might be interpreted as equivalent to describing one as more primitive than the other. From an evolutionary standpoint, it is hardly surprising that a more advanced fauna would succeed a more primitive one in time. But the two-phase kinetic model suggests more than a simple tautology. It suggests that the evolutionary dynamics of diversification into an essentially empty ecospace may make "primitive," generalized morphologies and ecologies advantageous, permitting groups of such character to radiate rapidly and to usurp the environment, thereby slowing the diversification of more advanced or specialized groups.

The picture is starting to come to life, but the models are still crude. Moreover, further study of the patterns of life as revealed at the family level was showing that something was badly wrong with the second half of Sepkoski's theorizing. Far from there being a steady plateau, achieved once and for all in the Paleozoic after the great extinction at the end of the Permian, the increase in diversity picks up again and, if we ignore human interference, continues full blast up to the present (Sepkoski 1984, 249).

The solution now, though, is obvious, especially since the (later) Paleozoic fauna seem to have been up to the same trick as the (earlier) Cambrian fauna—a peak of diversity and then a fading away as the third fauna, the Modern, picked up steam. One needs a third set of equations with corresponding curve to superimpose on the earlier two. This done, one gets the required picture (254). Sepkoski even claims that one can see something akin to the earlier miniequilibrium in the mid-Cambrian. At some point in the future, the rate of increase in diversity will cool off, and we will revert to a new (higher than before) state of equilibrium (see figures on pages 223 and 224; Sepkoski 1984, 262).

Sepkoski raised the question of whether the overall picture of diversity is approaching completeness. There seems not to be a fourth group of organisms lying in wait for the Modern plateau to be achieved. However, one cannot rule out the possibility of some kind of evolutionary innovation. After all, the plant kingdom came up with a fourth flora, the flowering plants (angiosperms), which did not appear until the Early Cretaceous. "Thus, by analogy to the plant record, we can speculate that one or more unpredictable innovations of importance comparable to

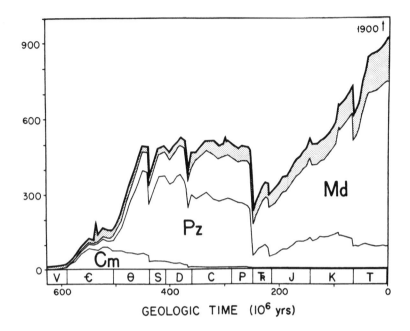

*The Phanerozoic history of the taxonomic diversity of marine animal families. The up-
per curve shows the total number of fossil families known to occur in each stratigraphic
stage of the Phanerozoic. The number "1900" is the approximate number of animal
families described from the modern oceans. The fossil diversity of these families is indi-
cated by the stippled field in the figure. The two curves below the stippled field divide
the diversity of heavily skeletonized families into three fields, representing the three
"evolutionary faunas" that dominate total diversity during successive intervals of the
Phanerozoic (from Sepkoski 1984).*

angiospermy might appear among future marine animals, leading to
major changes in faunal composition and driving diversity to yet higher
levels" (264).

Mass Extinction

The second major phenomenon crying out for Sepkoski's attention was
the massive and severe extinctions that seem to plague life forms at
certain points in history. The crash at the end of the Devonian, for
instance, and at the end of the Permian, and, famously, at the end of

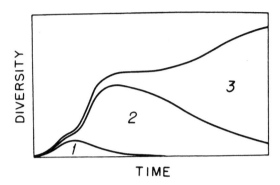

Sepkoski's computer modeling of the fossil record (from Sepkoski 1984).

the Cretaceous (when the dinosaurs went). Events like these seem to call for causal factors over and above those that bring on the constant extinction of species going on everyday.

Sepkoski worked on this problem with his colleague Raup, and in a now-notorious paper, published in the *Proceedings of the National Academy of Sciences* (thus by-passing the refereeing process, since Raup is a member of the Academy), the two paleontologists argued that not only is the geologic past marked by massive extinctions but over the past 250 million years these extinctions have come at regular intervals. They are periodic, with a mean interval of 26 million years (see figure opposite; Raup and Sepkoski 1984, 802).

Of course, the temptation to speculate on causes of such a finding is overwhelming, and Raup and Sepkoski gave in to temptation: "We favor extraterrestrial causes for the reason that purely biological or earthbound physical cycles seem incredible, where the cycles are of fixed length and measured on a time scale of tens of millions of years. By contrast, astronomical and astrophysical cycles of this order are plausible even though candidates for the particular cycle observed in the extinction data are few" (805).

This speculation was not done in isolation. Just a year or two previously, in 1980, the Nobel Laureate Luis Alvarez and associates had argued, based on geological evidence, that the extinction at the end of the Cretaceous which brought an end to the dinosaurs was no earth-caused event. Reasoning from a fine layer of iridium that was deposited

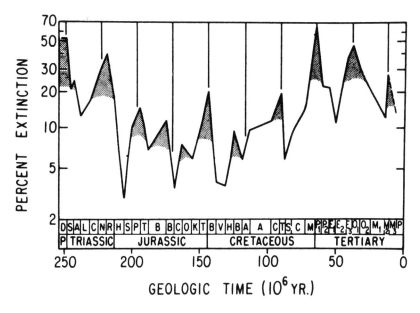

Extinction record for the past 250 million years (from Raup and Sepkoski 1984).

around the globe at the end of the Cretaceous, Alvarez and colleagues had speculated that the earth was hit by a very large comet or asteroid. This impact kicked up enough dust to cause a kind of nuclear winter that destroyed plant life and led to the starvation of the large denizens of the age. Although this hypothesis was controversial in 1980 (now it is widely accepted), it was plausible, and Raup and Sepkoski made explicit reference to it as a possible instance of just the kind of non-earth-originating factor they were supposing. "One possibility [of an extraterrestrial cause] is the passage of our solar system through the spiral arms of the Milky Way Galaxy, which has been estimated to occur on the order of 10^8 years."

As can be imagined, this thesis of periodic extinction brought on a massive negative reaction, almost equaling the force of the astronomical events about which the two paleontologists were speculating. Critics argued that their results are an artifact of the way we measure time in the past (Hoffman 1985, 1986; Kitchell and Estabrook 1986); of the ways in which the extinctions were measured (using families, and those not always distinguished in a fashion that would satisfy the ideologists of various taxonomic philosophies) (Patterson and Smith 1987); of the statistical methods used (reminding one of Disraeli's quip about lies, damned lies, and statistics) (Stigler and Wagner 1987, 1988); of insensitivity to background noise; and more (Maddox 1985; Hallam 1984). For some critics the extinctions do not exist, and for others they exist but are not periodic.

Raup and Sepkoski stuck to their guns (Sepkoski 1986a,b; Sepkoski and Raup 1986a,b; Raup and Sepkoski 1986, 1988). They claimed that a major reason why they went the path of publishing first in a nonrefereed journal was that they wanted immediate publication, since their thinking had become known to others. (*Science* had picked up and reported a talk on the subject given by Sepkoski.) But now, with the ideas officially in the public domain, they had the leisure to go by more regulated channels, and the periodic extinction hypothesis (which they could now apply to the more detailed level of genera as well as families) duly appeared in the refereed pages of *Science*.

In the authors' opinion, the case if anything was strengthened rather than weakened by this extension of the data to genera. And although they are prepared to make reservations—"The case for periodicity in the extinction record is based on statistical inference with messy data, and thus it cannot be proved or disproved in a truly satisfactory manner" (Raup and Sepkoski 1988, 96)—that is essentially how things have ended. Raup and Sepkoski see periodicity. Many others are still not convinced.

Worthy of note is the fact that, in their refereed discussions, neither Raup nor Sepkoski speculates about causes. Just after the reservation about the statistical nature of the evidence for their hypothesis, they admit that "acceptance of periodicity (and some of its suggested causes) would entail a major shift in the way geologists look at the history of the earth and of life" (96); but others were less cautious. News of Raup and Sepkoski's first paper circulated among astrophysicists, sparking causal hypotheses. A batch of these was published in *Nature*. One particularly eye-catching idea was that the sun has an unseen companion star which, when it comes close to our system, disturbs an (also unseen) comet cloud (Davis, Hut, and Muller 1984, 717; also Rampino and Stothers 1984; Schwartz and James 1984; Whitmire and Jackson 1984).

Raup took this interest as the spark to write a popular book on the whole subject, *The Nemesis Affair,* full of causal speculation. But in the professional refereed journals, the two paleontologists remained far more cautious, describing the evidence for periodic extinction but remaining largely silent about causes. Their caution was wise, for it has now been shown on theoretical grounds that the companion-star hypothesis is untenable: such a body would be unstable (Glen 1994). Today, Sepkoski inclines toward some sort of periodic comet shower: "You don't have to put in too much gunk between the earth and the sun to start upsetting the radiation balance and cut off the insulation budget by 1/2%. It would do some major things to the earth's climate system."

Insect Evolution

Although this controversy over periodic extinction attracted much attention, it was a minor external distraction from Sepkoski's ongoing personal research program. He has continued right on worrying about the nature and causes of diversity as revealed in the fossil record. Let me therefore bring this exposition to an end by referring to a recent paper on diversity, for once on the evolution of a specific group of organisms: the insects (Labandeira and Sepkoski 1993).

Many think that the insect fossil record must be sparse indeed, but this turns out not to be so. No less than 1,263 families are known from the record, going back to the Devonian. Their diversity in fact has almost always been greater than that of four-legged animals (tetrapods) (310). Although insects experienced high extinction rates during the Paleozoic, during the Mesozoic and Cenozoic insects have had much lower rates of extinction, leading to many similarities between past and present. The beetle genus *Tetraphalerus,* for example, has companion forms in 153-million-year-old Jurassic deposits. "Tetrapods, on the other hand, experienced major turnovers during the late Pliocene and latest Pleistocene . . . , and few living species are more than a million years old" (312).

Why the great success of insects, particularly since the Triassic? Could it be the appearance and diversity of the angiosperms? Sepkoski admits that "the great expansion of insect families toward the Recent would appear consistent with this proposition" (312). But so simple and straightforward an adaptive explanation does not appeal to Sepkoski. Indeed, he shows that the insects had been on their rise from the beginning of the Triassic: the arrival of the flowering plants altered things not one whit. "Whenever these plants originated, the fossil data indicate that angiosperms experienced a tremendous radiation in all geographic regions during the Albian and Cenomanian stages of the middle Cretaceous . . . However, there is no signature of this event in the family-level record of insects" (313).

Sepkoski shows that the insects had in fact evolved the kinds of adaptations needed to exploit the angiosperms before these plants actually arrived. In a sense, therefore, they were pre-adapted. But the main point—the leitmotif of Sepkoski's work—is that the significant patterns in the fossil record are not to be tied directly to adaptive advantage. They are rather to be found in the interplay of large numbers, dependent on processes of origination and extinction of entire taxa. It is not so much that Sepkoski is denying adaptation at the individual level but rather that—here as before—he does not see its direct relevance to the overall pictures that he finds in the fossil record.

Values in Sepkoski's Science

Sepkoski belongs to a generation of paleontologists self-consciously trying to model their practices on those found in the rest of biology. In his early paper on computer-generated evolution—whose coauthors included Gould, Raup, and Simberloff (Gould et al. 1977, 24–25)—one learns:

> We believe that paleontology—the most inductive and historical of sciences—might profit by applying some deductive methods commonly used in the non-historical sciences (without sacrificing its important documentary role for the history of life). We may seek an abstract, timeless generality behind the manifest and undeniable uniqueness of life and its history. We take as our guide the recent success of simple, general models in the other branch of natural history most celebrated for the complexity and uniqueness of its subject—ecology.

MacArthur and Wilson's work on island biogeography is referenced explicitly as one of their models. We should therefore expect to find that already-encountered epistemic norms are important in Sepkoski's work—and they are.

Obviously a paleontologist cannot make straightforward predictions: by definition, his or her subject matter is dead! But if we think (as we should) of prediction as being a logical inference from the known to the unknown, then everything that Sepkoski does is dedicated to making a predictive science of his subject. Take the initial work on Phanerozoic taxonomic diversity. Sepkoski starts with certain initial conditions —organisms under constraints of origination and extinction—and then shows how from these conditions or premises the known pattern of the fossil record follows. Granted, a pragmatic shuffling of theory and data was necessary to bring the two into harmony. But that is the nature of science. Real science is always more like auto mechanics—getting the damned thing to work—than is dreamed of by philosophers in their texts on scientific methodology. The point is that Sepkoski did get the damned thing to work: rather well, in fact. (One critic was Hoffman 1989, who is answered effectively in Sepkoski 1991.)

More than this, as his ongoing work shows, Sepkoski then kept tinkering—revising and augmenting—as more and better-refined data were obtained. As he moved from orders to families, fresh items emerged from the fossil record, and the same is true as he (and others) gathered more information overall. The second half of the Phanerozoic, for instance, was seen not to be in equilibrium, and this required more sophisticated and complex modeling than before—which Sepkoski provided.

Sepkoski is self-conscious about the desirability of having a predictive science. This is one of the reasons why he continues to defend his extinction periodicity hypothesis. Originally, two peaks (signifying extinction) were missing from the pattern:

> This is when I started collection data on genera, because I knew families were insensitive to small gaps, and the first thing to drop out of the genus data was to fill in one of those gaps—right where we predicted it would be. Nothing else came that we hadn't seen

in the family data. This is one event right where we predicted. A few years later, after cleaning up the data—collecting more and so on and so forth—a second peak starts to emerge, right where we had predicted it would be. No other peaks are emerging in this data, that we didn't know about beforehand. So, you know, paleontology isn't known as a predictive science. This is what predictive science is supposed to do. Tell you what's missing and make you go look to fill it in, because you know where to look now. And this is exactly the periodic pattern. And so, if that's not good corroboration, I don't know what is.

Moving on to other epistemic norms, we find that they too get attention. Clearly something like internal consistency or coherence is important. The diversity of the Archaeocyathids fails to fit the expected pattern of Cambrian diversity growth (Sepkoski 1979, 229). Perhaps, suggests Sepkoski, the problem is with the Archaeocyathids rather than the model: for instance, taxonomists may have been subdividing them more than is warranted, giving an illusion of undue diversity, or they may not be true metazoa at all. Arguments based on independent evidence are offered as to why both of these possibilities may be true. Hence, an apparent weakness in Sepkoski's theorizing is turned into strength. Rather than an unacceptable hypothesis, we may have a strongly confirming unexpected prediction.

The need for external consistency is shown by the appeal to the astrophysicists for help with the causes of the supposed periodicity of mass extinctions. If one is going to rely on extraterrestrial causes, then one had better have one's physics right—and who better to get the physics right than physicists themselves? Unification and simplicity, although they are hardly paraded publicly, seem also to be important factors. Certainly one can say that Sepkoski's work on the causes of divergence in the fossil record are bound together by a remarkably simple—elegant, not simple-minded—set of assumptions bound into a very few models.

And so to predictive fertility. A neat example showing how Sepkoski takes this norm seriously occurs in a paper he coauthored on the mammalian interchange between the North and South Americas which occurred when they were first joined a mere 3 million years ago

(Marshall et al., 1982). What one finds is that initially the numbers of taxa (families in this study) rose from 32 to 39 in the South, and comparably in the North, and then fell back somewhat to 35 in the South, and again comparably in the North. Out once again comes a paleontological equivalent of the MacArthur-Wilson biogeography theory, and it is shown how this can be applied to the Americas' interchange.

At first one gets an increase in numbers and then a decline, powered especially by a decline in the numbers of the original taxa. "If taxa immigrate into a large area containing a native fauna, such as occurred in South America, the addition of immigrant taxa will, in essence, supersaturate the fauna of the new area. Extinction rates of both native and immigrant taxa will increase as diversity exceeds the equilibria of both faunal components. This will slow the increase in immigrant diversity and cause an exponential decline in native diversity" (1355). Thus the equilibrium model of island biogeography is extended to new fields.

Sepkoski takes epistemic norms very seriously indeed. In this respect, he differs little if at all from Parker—a comparison which extends to a shared comfort with mathematical techniques. But Sepkoski does differ very greatly from Parker when it comes to nonepistemic factors, particularly the presence or absence of adaptationist tendencies. This has little to do directly with the fact that the one is a neontologist, working on the behaviors of living organisms, and the other a paleontologist, trying to work out patterns of life among long-dead organisms. A paleontologist can be an ardent Darwinian, trying to discern and understand instances of adaptation. It is just that this is not Sepkoski's way. Although his background is Polish Catholic, other than a morbid fascination with the politics of the Church, his general attitude to Christianity has long been one of indifference. At high school, "the Christian Brothers had beaten religion out of me while trying to beat it into me." So adaptationism was not making an entry in this direction.

Neither was it making an entry in science. As an undergraduate at Notre Dame, Sepkoski took no courses at all in biology, and by the time he got to Harvard what really excited him were computers and programming and so forth. He wanted science where he could use this expertise. "Despite taking courses with Steve [Gould] on allometry and morphological analysis and so on—it just never really appealed to me." Little wonder that the side of evolution dealing with adaptation left Sepkoski cold. Even the sociobiology debate between former teachers failed to excite: "I've never been interested particularly in the evolution of behavior. And so I wasn't particularly interested in the details of the debate."

But do not think that this means Sepkoski stands outside his culture. Sepkoski's evolutionism is as American as it is possible to be. Notwithstanding the fact that Sepkoski has never in his life read a word of Herbert Spencer, the legacy—remember, the influence was far greater in North America than in England—persists. Go back to Sepkoski's picture of life during the Phanerozoic. It is pure Spencer. We climb up to a plateau—a plateau that Sepkoski refers to as a state of kinetic or "dynamic equilibrium"—and then something else takes over and we have a new climb up to another plateau. The student of Wilson and of Gould had learned his lesson well. Rise and pause; rise and pause.

But could you not say that, rather than a consequence of a particular cultural view of the world, this is just a description of the evidence and not even a particularly theory-laden one at that? This is the interpretation that Sepkoski himself would put on his work: "If you go through those kinetic model papers, they're really quite descriptive, they're quite empirical. And the axe I'm grinding is that diversity looks like it's an equilibrium process . . . And then I'm simply using the data to look at it." But, with all due respect, we are being offered no naive empirical interpretation here. The picture was produced by cutting and shaping in the light of theory. I do not in any sense see this as dishonest or falling from the standards of good science—far from it—but it does make a mockery of claims for a disinterested, uninterpreted picture.

More than this, the picture is incomplete. Thus far, the major plateau at the end of the third (present-day) phase is nonexistent! It is supposed—extrapolated—rather than seen. There is nothing wrong with this (although given human interference, we may never get the plateau); but again we have reason to think that we have a culture-drenched vision, a neo-Spencerian vision, rather than raw data.

On top of all of this, the way in which diversity is supposed to build is more Spencerian than Darwinian. Not in some obvious respects, of course. Spencer was a Lamarckian and Sepkoski is not. But in the belief that selection, although perhaps significant (which Spencer too would have allowed), is not really the driving force of evolution. Life's history is no random woodpile, thrown together by the messy contingency of natural selection. Instead, for Sepkoski there is an almost cosmic inevitability about the course of evolution. It has built-in parameters; and, once started, not even mass extinctions can deflect it from its course. Remember also that for Spencer, the upward rise is intimately connected to the fact that fertility starts to drop away: the high-grade mammal has far fewer offspring than the low-grade fish. Likewise in Sepkoski (1984)—although with the Gouldian twist of thinking at the group rather than the individual level—we see the same intimate connection with fertility: a falling away of the rate at which new taxa are produced. All yielding a progressive life history, as the later more successful, more specialized organisms of the Mesozoic and Cenozoic replace the earlier, less successful, more generalized organisms of the Paleozoic. Homogeneity into heterogeneity by another name.

Are values involved here or is Sepkoski (unconsciously) simply reflecting an American tradition, passed down from Wilson and going back to Spencer's synthetic philosophy? Precisely because his field (paleontology) has a bad reputation to counter, Sepkoski is supersensitive to the directive that professional science be value-free. "I have my interests, but they don't necessarily get intertwined with my science." In fact, Sepkoski is not above criticizing his mentors for sloppy habits which he fears will introduce cultural values into science:

> I actually don't like the use of metaphor in science except in very restricted ways. I think Steve Gould and Dick Lewontin are at fault for depending too heavily on metaphor in science. Metaphor is not an exact way of expressing a scientific concept. In fact, it's often almost like cheating, a cheap way of trying to express a concept. Sometimes metaphors encapsulate something, but one can get away from rigor by depending too heavily on metaphor. So, in trying to eschew metaphors in my work, I don't bring politics, religion, and so on, into my science.

Sepkoski is aware that some kind of metavalue influences his work; but again he would deny that this puts the wrong sorts of cultural values into his science. "I'm not a conservative person and my science has not been conservative. It has not been following the well-worn path of Kuhnian normal science. I've had more fun exploring the new ways of looking at things. The new ways of describing patterns in the history of life. This isn't political or religious but more a world view. I'm interested in dynamic aspects of large-scale evolution." To this we might add a cherishing of science which lends itself to a quantitative approach, especially one open to computer-driven research. Making massive data collections and grinding them through programs is as culture-bound as is the heavy use of game theory.

But what about progress? If Sepkoski is as Spencerian as I claim, surely the old Englishman's values must show through sometimes? Sepkoski is at pains to deny this. He points out that "general" may well be the preferred option when an ecological space is empty, whereas "specialized" is the preferred option when species must struggle for living room. In his own words:

> I'm just trying to see how one evolutionary fauna can replace another in terms of usurping the environment—members of the later fauna usurping the environment or ecospace of the preceding fauna. And if there's more of the succeeding than more of the preceding, perhaps it's because they're subdividing the environment so you can push more in and so on and so forth. But, I read it so, rather than in terms of the larger question of progress or complexity or anything like that.

Perhaps if you keep pressing long enough and hard enough, some values do emerge. Consider the following passage from an interview given at the time of the extinction controversy (winter 1989):

> Mass extinctions have probably been good for the evolving bio-sphere. I said "good" and I've got to explain why I said "good"—in the sense that they've probably promoted diversity. Real evolu-tionary innovations, probably coming in during the rebound of these extinction events, clear out a lot of diversity. Clear out a lot of biomass. We're back into semifrontier days. Sort of environ-

ment where you don't have to be real good to get on, so something very new and different may be able to grab hold of a piece of the ecological pie and hold it, giving rise to new kinds of organisms. So mass extinctions are good in that sense. They promoted evolutionary innovation.

At this point the sympathetic reader may agree that if this is the best that can be said, then even if values seep through, the damp can quickly be mopped up. The stain is hardly troublesome and certainly not ruinous. Sepkoski is much aware of the sociological imperative that good professional science be cultural-value-free. Although his science is certainly influenced by his culture as such, this is not to say that it is value-impregnated in any significant way.

Evolutionism as Professional Science

On the surface, Geoff Parker and Jack Sepkoski could not be more different. The English Darwinian sociobiologist versus the American Spencerian paleontologist. But on the value scale which judges between the epistemic and nonepistemic, they are virtually at the same point. They both take epistemic values very seriously, self-consciously so. They are both influenced by culture. Whether or not this ever happens in their own science, they are both very wary about the intrusion of nonepistemic values into science. And although both have been in the public eye, ultimately both feel uncomfortable about letting professional science descend into the public arena. Sepkoski is almost censorious about the very existence of the sociobiological controversy. "It sounded very politically motivated and nothing to me. And so I didn't become terribly involved in the whole thing." One can hear echoes of Parker.

On this note, with our history now completed, let us turn our full attention to analysis and see what we have learned about the nature of science. From primitive beginnings, we have now professional science: and it shows in the work being produced. Let us see what this signifies.

12

❦

Can Evolution Cut the Mustard?

We have arrived at the end of our history of evolutionary thinking. The time has come to look again at the question that set us off through two and a half centuries of speculation on the origins of organisms. What is the true nature of science? Is it objective? Is it, as Karl Popper said, "knowledge without a knower"? Something which tells us about the real world, out there? Or is it subjective, as Thomas Kuhn and those following him have suggested? Is science a reflection or epiphenomenon of culture? Something which changes as society changes and which tells us less about reality and more about ourselves? Is science a social construction? Is evolution a social construction?

Start with the history. Our first finding is that, on the epistemic front, objectivists are surely right. There is a set or body of norms or values or constraints that guides scientists in their theorizing and observing: predictive accuracy, internal coherence, external consistency with the rest of science, unificatory power (consilience), predictive fertility, and to some degree simplicity or elegance. Satisfying this set of demands is the mark of good science—the kind of science one expects from a professional. And while one person or group may tend to stress one or more values rather than another, with respect to our study there is something transcultural about them. They are above the vagaries of societal change or whim or fashion. In this sense, they are pointing toward truths about a real world: objectivity.

The second finding is that the history of evolutionism, from the middle of the eighteenth century to the end of the twentieth, is one of ever-greater manifestation and adherence to the epistemic norms. Erasmus Darwin had but a casual relationship with the standards of good science: his work was hardly predictive or anything else much. Charles Darwin took magnificent leaps forward. At the same time he had epistemic weaknesses: there was certainly nothing much by way of exact prediction, and there were perceived epistemic failures as well—the inconsistency of his theory with the earth-dating findings from physics, for instance. Moving forward to this century, we see yet greater attempts to make evolutionary theorizing epistemically rigorous. First, through the work of people like Theodosius Dobzhansky, and more recently through the next generations: Lewontin and Wilson, Parker and Sepkoski. As we prepare to leave this century and move into the next, we can truly say that the epistemic norms play a major role in the structure of evolutionary theorizing and that their satisfaction is significantly above what it was in earlier times.

The third finding is that cultural values were important—all important—at the beginning, and that within science we have seen a gradual diminution or restriction of their importance. For Erasmus Darwin, the value of social or cultural progress was the very reason for his evolutionism: biological upward movement was a philosophical theory made flesh, one might say. For Charles Darwin things were very different. One simply cannot say that his evolutionism was no more than an excuse for values that he held dear. Epistemic virtues like consilience were at the heart of his thinking. Yet, although more cautious, he certainly did not see the nonepistemic or cultural as entirely incompatible with his science: remember his enthusiasm not only for progress but for other of Victorian Britain's social, racial, and sexual divisions. The cultural influence lay heavily on evolutionary theory long after Darwin, especially in the theory's Spencerian incarnation. However, in this century, the cultural declines yet further. Although we have people who specialize in popular-science-level evolutionary theorizing, today the most professional of modern scientists spurn the cultural values within science.

The Metavalues of Evolution

Connected with the expulsion of the cultural, nevertheless, we have metavalues—values that are about science rather than within it. Most particularly, we have the metavalue which asserts that good science should not have any cultural-value components. I do not see this as entirely innocent: in the early nineteenth century Cuvier condemned evolutionary thought as culturally value-laden because he (a modestly born Protestant) wanted to convince his conservative Catholic masters that his own work as a scientist was ideologically safe. In this century Dobzhansky strove to make his professional work value-free, else he would not get any of the grants being offered. But the effect is to push overt cultural values out of professional science.

Julian Huxley came to naught in major part because he would not obey this metavalue. Edward O. Wilson has skated dangerously close to the edge on occasion because he too has been cavalier with its demands. Whether cultural values have been eliminated from science intentionally or whether (as I suspect) this metavalue has now been unconsciously internalized as part of the culture of science, the fact is that modern evolutionary biology frowns strongly on the intrusion of cultural values into what is intended as professional—the best-quality—science.

All in all, at a preliminary level, the history of evolutionary theory shows a move—a strong and decisive move—from the subjective to the objective (as judged in terms of the epistemic/nonepistemic dichotomy). Nor is this a conclusion negated by the metavalues that influence the epistemic standards scientists employ. It is true that all of these directives stem from cultural values in one sense or another, and they certainly affect the finished scientific product. But overall these metavalues do not compromise the objectivity of science—the very opposite, in fact. Generally the end of the metavalues is to promote the very things that lead us to think of science as objective.

The Metaphors of Evolution

Although cultural values may have declined in evolutionary biology, in our evolutionizing today we still rely very heavily on many, many elements from our culture and that of our forefathers. A list comes readily to mind: struggle for existence, natural selection, sexual selection, adaptive landscape, dynamic equilibrium, arms race, and much more. These do not even start to include the more controversial and arguable popular notions, like selfish genes.

My point is that there is still something deeply cultural about evolutionary biology, even at its most mature or professional or praiseworthy level. Through the language, the ideas, the pictures, the models, above all the metaphors that evolutionary biology uses, culture comes rushing right back in. This was true at the time of Charles Darwin and it remains true at the time of Geoffrey Parker and Jack Sepkoski. From the tree of life to evolutionarily stable strategies, we have culturally rooted metaphors: an idea from one domain, that of culture, is taken and applied to another domain, that of organisms (Hyman 1962; Beer 1983; Myers 1990; Selzer 1993). Progress is obviously a metaphor starting with culture, and indeed the same is true of evolution itself: it was a notion originally dealing with the development of the individual and only gradually applied to the development of species, and has ever been surrounded by all kinds of thoughts about divine creation and so forth (Richards 1992).

In speaking of cultural metaphors I am not now talking of values by another name, or by no name at all. I have no reason to think, for instance, that when the white southerner Wilson talks of slave species of ants, he is thereby showing solidarity with the antebellum South. I would certainly agree that cultural values can seep back into science because of the use of cultural metaphors. It is difficult to think of evolutionary trees without getting progressionist, precisely because trees in our culture are associated with upward striving. The metaphors of up are positive ("Stand up! Stand up, for Jesus!"), whereas the metaphors of down are negative ("Are we downhearted? No!") (Lakoff and Johnson 1980). But values are not my point here. I am simply claiming that when evolutionary scientists turn to language to express their findings, the words they choose are often laden with metaphors taken from the surrounding culture.

Not such a simple claim, of course. Even without values, opening the door to culture seems to invite subjectivity and relativism in. Not only do we have to contend with the fact that Ed Wilson lets his epistemic theorizing be influenced by his childhood in a militaristic society, but we must also contend with the fact that Wilson thinks adaptively because we live in a society which (probably in major part because of our Christian heritage) thinks in terms of function for organisms. And that he argues in terms of equilibrium because of American traditions which go back to the influence of Herbert Spencer (Russett 1966, 1976). That he puts everything down to natural selection because we today are reaping the benefits of the agricultural advances brought about by the artificial selection of the early nineteenth century (Kimmelman 1987). And that he believes in evolutionary trees because he does not live on the Canadian tundra. Without any of these ideas, Wilson would be no more than a graduate student in search of a thesis topic. Without the metaphors of his society, his science would not exist.

Which all seems to imply that at a different time and in a different place, Wilsonian science would have been different. Not necessarily better, but different. And different in ways that reflect cultural differences. Such a conclusion is not quite as bad as saying that Wilson's science is simply a figment of his imagination—an exercise in wish fulfillment or an extended polemic about the way he would like things to be. But not a great deal better. It still says that science is a reflection of society rather than the real world. It is in this sense deeply subjective. Science is indeed a social construction, and the fact that we may like the end product does not make it less relative. Kuhn himself suggests that metaphor is a key element of the picture he tries to capture through his notion of a paradigm. This suggestion has come back to haunt us.

Something seems to have gone wrong with the objectivist case. So badly and so obviously wrong that one suspects a flaw in the argument just given. And indeed, objectivists will have little trouble finding it. Since we have focused on metaphor as a major (if not the exclusive) way in which culture insinuates itself into evolutionary biology, let us stay for a moment with that. The point which should be obvious to anyone (says the objectivist) is that, although metaphor is extremely widespread in human discourse, it is not essential. It is in a sense—an important theoretical and perhaps practical sense—eliminable. We all use metaphors, but they are in the last resort short-hand for literal language. Nothing which is said through metaphor cannot be said otherwise, in plain terms which speak directly to reality. Wilson's science may use metaphors, but they could be removed. And then we would be left with a fully objective (and perfectly adequate) residue. "When you actually start to do the science, the metaphors drop out and the statistics take over" (Fodor 1996, 20).

In making this argument, the objectivist is appealing to a grand old tradition. It goes back to Aristotle (*Poetics* and *Rhetoric*), so let us treat it with respect. But this does not mean that we must accept it. The point I would make—one which we have seen exemplified in our history—is that even if metaphor could in theory be eliminated, no sensible scientist would ever think seriously of making such a move. Most crucially, one would at once lose one of the most important of the epistemic values, namely, predictive fertility. Metaphors, as it is sometimes said, are absolutely vital for their "positive heuristic" as they push one into new fields and new forms of thinking (Hesse 1966). Without metaphors—which are vehicles for seeing similarities in otherwise dissimilar things—one would lose a value so essential that in its absence science would simply grind to a halt. At best, one could play variations on themes already heard. And as one repeats oneself endlessly, hopes of fresh triumphs in other epistemic directions—predictive accuracy, unificatory power, compulsive elegance—would get lost also. To requote McMullin (1983, 16), the best sciences "have the imaginative resources, functioning here rather as a metaphor might in literature, to enable anomalies to be overcome and new and powerful extensions to be made."

Let me make the point through a heuristically crucial metaphor we encountered in Wilson's work. I refer to the division of labor. Wilson did not of course invent this notion himself. It goes back to the eighteenth century, getting detailed treatment in Adam Smith's *Wealth of Nations*. The idea of assigning to each worker his or her own allotted task is a key element in the smooth functioning of industry, as important in its way as any new piece of machinery. It is as much part of the culture of the age as the spinning jenny or, later, the steam engine. And naturally it was tied into progress, for a society utilizing the division of labor to the full was seen as much advanced over one where it was unknown or little used.

The idea can be found in the evolutionary writings of Erasmus Darwin, but it is really introduced into biology by the Belgian-born

French biologist Henri Milne-Edwards (1827, 1834), who spoke of a physiological division of labor, meaning that the different parts of the body are specialized for different tasks. It was picked up by that energetic intellectual scavenger, Charles Darwin—he acknowledged the influence of Milne-Edwards but surely also his family background was a factor (his maternal grandfather, Josiah Wedgwood, had made a fortune from the successful application of the division of labor)—and Darwin used it repeatedly in different ways.

Right off it guided his thinking about barnacles, the massive classification of which was his occupation in the late 1840s and early 1850s. Following Milne-Edwards in believing that the division of labor spells progress in the world of biology as much as in the social world, Darwin used the notion to that end when he decided where on the scale of nature the barnacles should be placed: "Barnacles in some sense, eyes & locomotion, are lower, but then so much more complicated, that they may be considered as higher" (quoted in Richmond 1988, 392). Later, when he could be explicit about his evolutionism, he wrote in discussion of the reasons for barnacle sexuality that "a division of physiological labour is an advantage to all organisms" (Darwin [1873] 1977, 2, 180).

Then, when Darwin came to the *Origin,* the metaphor of a division of labor is first tied into the so-called principle of divergence, crucial inasmuch as it explains (and thus can predict) why a group of organisms might split into two or more descendant groups (115–116):

> The advantage of diversification in the inhabitants of the same region is, in fact, the same as that of the physiological division of labour in the organs of the same individual body—a subject so well elucidated by Milne Edwards. No physiologist doubts that a stomach by being adapted to digest vegetable matter alone, or flesh alone, draws most nutriment from these substances. So in the general economy of any land, the more widely and perfectly the animals and plants are diversified for different habits of life, so will a greater number of individuals be capable of there supporting themselves.

Later in the *Origin,* the division of labor reappears to explain the existence of more than one form of sterile ant in a colony. "We can see how useful their production may have been to a social community of insects, on the same principle that the division of labor is useful to civilised man" (241–242). Not that Darwin had anything against applying the division of labor to the parts of the individual. This idea occurs in the *Origin,* most particularly in the third edition when Darwin is explaining why he thinks humans to be superior to other organisms. Here Karl Ernst von Baer gets credit for having enunciated the significance of "the completeness of the division of physiological labour" (Darwin 1959, 221).

And then, once one gets to *The Descent of Man,* the individual application of the division is a key theme. Referring to the human ability to walk upright, Darwin wrote: "To gain this great advantage, the feet have been rendered flat, and the great toe peculiarly modified, though this has entailed the loss of the power of prehension. It accords with the principle of the division of physiological labour, which prevails throughout the animal kingdom, that as the hands became perfected for prehension, the feet should have become perfected for support and locomotion" (Darwin 1871, 1, 141–142).

Nor was the group concept of the division of labor missing from *Descent.* In talking of social evolution, Darwin wrote that early man divides his labor, as he molds his ways to his changing conditions: "He invents weapons, tools and various stratagems, by which he procures food and defends himself. When he migrates into a colder climate he uses clothes, builds sheds, and makes fires; and, by the aid of fire, cooks food otherwise indigestible. He aids his fellow-men in many ways, and anticipates future events. Even at a remote period he practiced some subdivision of labour" (1, 158).

Before, during, and after the *Origin,* the sociocultural idea of a division of labor was transferred right into the evolutionary thought of Charles Darwin. Yet in all of the uses, in some way it is not just the nonepistemic idea which is being endorsed and promoted. Darwin does not just break off discussion to talk in isolation of the virtues of the division of labor. Nor does he praise the division at the expense of the epistemic power of his theorizing. In fact it is not because it is a valued concept that it is important. Rather, Darwin is using the cultural concept (irrespective of his feelings of its worth) to further his epistemic ends. He can, for instance, predict what will happen when a group is faced by different open ecological niches (Darwin 1859, 116):

> A set of animals, with their organisation but little diversified, could hardly compete with a set more perfectly diversified in structure. It may be doubted, for instance, whether the Australian marsupials, which are divided into groups differing but little from each other, and feebly representing, as Mr Waterhouse and others have remarked, our carnivorous, ruminant, and rodent mammals, could successfully compete with these well-pronounced orders. In the Australian mammals, we see the process of diversification in an early and incomplete stage of development.

In a similar manner, Darwin can bring the study of the social instincts into the consilience-exhibiting, unified evolutionary family; he can explain anatomy; he can push his ideas in a fertile manner into new dimensions of human behavior and evolution; and much more. In this sense, we get a meshing of the epistemic and the nonepistemic: culture promoting the epistemic. We have a blend like water and alcohol, rather than a mix like water and oil.

Darwin was not the only Victorian evolutionist to make much of the division of labor. It was, if anything, even more significant in the thought of Herbert Spencer (1862). He identified progress with the move from homogeneity to heterogeneity, in culture and in biology. For him, therefore, movement toward a division of labor was progress by another name. And so it was to continue, down to the age of Edward O. Wilson, for whom the division of labor is a crucial tool for thinking about the social insects. Remember how his ant genus *Atta* made that vital move that we humans were to make many millions of years later: "The fungus-growing ants of the tribe Attini are of exceptional interest because, to cite the familiar metaphor, they alone among the ants have achieved the transition from a hunter-gatherer to an agricultural existence" (Wilson 1980a, 153). And this has meant a division of labor: the queen, the larvae ("which may serve some as yet unknown trophic function"), and some seven castes of workers in all (150):

> A key feature of *Atta* social life disclosed by these data is the close association of both polymorphism and polyethism with the utili-

zation of fresh vegetation in fungus gardening . . An additional
but closely related major feature is the "assembly-line" processing
of the vegetation, in which the medias cut the vegetation and then
one group of ever smaller workers after another takes the material
through a complete processing until, in the form of 2-mm-wide
fragments of thoroughly chewed particles, it is inserted into the
garden and sown with hyphae.

Then follows the work of seeing how all of this might have evolved
and the role played by natural selection and if (as Wilson thinks) the
adaptations as manifested through the division have been "optimized"
and so forth: basically the ants do not use either those at the bottom or
at the top limits of possible leaf cutters. "What *A. sexdens* has done is to
commit the size classes that are energetically the most efficient, by both
the criterion of the cost of construction of new workers . . . and the
criterion of the cost of maintenance of workers" (Wilson 1980b,
163–164).

The metaphor of the division of labor structures Wilson's discussion and leads to the answers. My suspicion is that for Wilson—as for
Darwin and Spencer—the division of labor is not entirely a value-free
notion. Given Wilson's progressionist sentiments, one certainly suspects that he views it with a favor that transcends the epistemic. But the
value side is diminished and there is nothing in Wilson's work—or
much of Darwin's work, for that matter—which necessitates a cherishing of the division of labor. One could be indifferent to it in human
terms—disliking it even and trying to transcend it (as many modern
industries now attempt to do)—and still make full use of it in one's
evolutionary theorizing. There is no reason why ants or body parts
should find specialization deadening or degrading in the ways that can
be true of humans. For Wilson (1980a), the division of labor as incorporated in the caste system is "an essential part of the specialization on
fresh vegetation" and, conversely, the way in which fresh vegetation is
used "is the raison d'être of the caste system and division of labor" (150).
But also for Wilson, as for Darwin, the idea of a division of labor, rooted
in modern Western culture, is a key to the epistemic success of his
science. In this sense, I see the influence of culture on scientific ideas as
something that is here to stay.

Anything Goes?

Although I deny that culture necessarily imports a value component into science, I argue that it has an essential role. Does this mean that science is simply a slave to culture? As goes culture, so must go science? Not at all! Rather than precluding the satisfaction of epistemic norms, culture makes them possible. At this level, objectivity—respect for and satisfaction of epistemic standards—floods back in. Think of something like Popperian falsifiability—the flip side to the demand that good science be predictively accurate. Through the metaphors of culture, predictions are made possible. But then the science produced can and must be judged simply by the epistemic standard of empirical success. However socially or culturally congenial one may find the science, if it does not succeed in the fiery pit of experience, it can and should be rejected.

Take for example the question of the level at which selection is supposed to operate. Cultural factors have played a crucial role in the positions that people have taken. Charles Darwin was influenced deeply by the thinking of eighteenth- and nineteenth-century political economy, especially those aspects that were congenial to the successful industrialists. For him, ultimately, the struggle always pits individual against individual (Ruse 1980). Hence, adaptations are always for the benefit of the individual. They are in this sense "selfish," to use Dawkins's metaphor. Others have seen the struggle differently, as occurring much more between groups than between individuals. And there were still others for whom the whole idea of struggle is alien. In both cases the Darwinian position is denied.

A. R. Wallace saw the struggle as occurring between groups. He was much influenced by his early experiences as a land surveyor: he saw conflicts between the social classes, as those in power enclosed and took away the land from those beneath them. Combined with a powerfully favorable exposure to the socialistic teachings of Robert Owen, he had always a tendency to see genuine alliances within groups and conflict between them (Wallace 1905). Likewise, Darwin's great German supporter, Ernst Haeckel, saw conflict between groups—no doubt a function of the conflicts between the Prussian state, wherein he was a professor, and the recently conquered France. And he too saw cooperation within the groups—again perhaps a function of the Prussian state with its efficient civil service and its high-quality, state-supported education (Haeckel 1866, 1868).

Russian evolutionists in the nineteenth century denied the interorganic struggle for existence altogether. Russia was late in industrializing: indeed, one might fairly say that it never truly did industrialize under the tsarist regime. The whole Adam Smith/Robert Malthus tradition was alien and irrelevant to the Russian experience (Todes 1989). One looked for other philosophies like socialism or, in the case of Prince Petr Kropotkin, anarchism. As important, Russia was so vast and with a climate so cruel that no one could ever think that a struggle between organisms was a significant factor. There was always going to be space enough. The key struggle was between organisms and the elements (Todes 1989, 128–129, quoting Kropotkin 1902, vi–viii):

> The terrible snow-storms which sweep over the northern portion of Eurasia in the later part of the winter, and the glazed frost that often follows them; the frosts and the snow-storms which return every year in the second half of May, when the trees are already in full blossom and insect life swarms everywhere; the early frosts and, occasionally, the heavy snowfalls in July and August, which suddenly destroy myriads of insects, as well as the second broods of birds in the prairies; the torrential rains, due to the monsoons, which fall in more temperate regions in August and September—resulting in inundations on a scale which is only known in America and in Eastern Asia, and swamping, on the plateaus, areas as wide as European States; and finally, the heavy snowfalls, early in October, which eventually render a territory as large as France and Germany, absolutely impracticable for ruminants, and destroy them by the thousand—these were the conditions under which I saw animal life struggling in Northern Asia. They made me realize at an early date the overwhelming importance in Nature of what Darwin described as "the natural checks to overmultiplication," in comparison to the struggle between individuals of the same species for the means of subsistence.

Clearly the only way that people—or organisms—could survive was by banding together against the elements. It was not by chance that

Kropotkin, living in exile in London, penned the greatest-ever paean to a natural form of altruism, mutual aid (Todes 1989, 134, quoting Kropotkin 1902, 293):

> In the animal world we have seen that the vast majority of species live in societies and that they find in association the best arms for the struggle for life: understood, of course, in its wide Darwinian sense—not as a struggle for the sheer means of existence, but as a struggle against all natural conditions unfavourable to the species. The animal species, in which individual struggle has been reduced to its narrowest limits, and the practice of mutual aid has attained the greatest development, are invariably the most numerous, the most prosperous, and the most open to further progress . . . The unsociable species, on the contrary, are doomed to decay.

Kropotkin was not peculiar in this. Indeed, he stood firmly in the Russian tradition.

Here we have rival theories representing rival views of the world. Different metaphors have been incorporated into evolutionary theorizing. On the one side, we have the metaphors of British industrialism. On the other side, we have the metaphors of socialism, bureaucracy, and a peculiarly Russian experience. But however deeply the culture may run, scientists do have rules of proper scientific conduct which they share with those scientists of other cultures—rules of conduct incorporated in the epistemic values. And this means that the two perspectives—the individual selection and the group selection views—can be compared. As indeed they were. And, notwithstanding some technical exceptions and despite some antediluvian holdouts, the one was found satisfactory and the other lacking. Most importantly, the individual selection hypothesis has been found predictively fertile in ways that the group selection hypothesis simply is not. Geoffrey Parker's life work is a testament to this fact. From a group perspective, none of the actions of his dung flies makes sense. Why should one male fight to overcome another, or why should any male give up copulating before all of the eggs are fertilized? From the perspective of the individual, all of these things are made clear, and quantifiable predictions are possible and made and confirmed.

Parker is the brilliant tip to a very large iceberg. Although culture may throw up different metaphors leading to different scientific approaches to a problem (not to mention to different problems), this does not mean that we are plunged into crippling subjectivity, where one simply cannot compare the different approaches and where—in the interests of moral or social purity—we have to respect all approaches equally so as not to show insensitivity to human diversity. If an approach, however sincere, does not cut the epistemological mustard, then it must go—as has been the fate of traditional group selection. Culture is important, but there are standards, those expressed by the epistemic norms. And they are indifferent to race, sex, class, and cultural heritage.

In short, for all they may engage in metaphor at a heuristic stage of theory development, scientists can and must go out and, in good Popperian fashion, test their ideas against experience. According to their results, they may then retain or they must modify or reject their theories, as well as their metaphors. So, along with subjectivity there is an objective element to science. Complementing the Kuhnian spin on science conferred by the metaphors of culture is a Popperian dimension. Both of our philosophers captured part of the overall picture.

Realism or Nonrealism?

The time has come to draw together the threads. But how can we think of doing this with the most important question of them all as yet unmentioned, let alone solved? Our ultimate concern is surely with the issue of realism. Does an objective, "real world" exist "out there" that can be known through the methods of science, or is science a subjective construction corresponding to shifting contingencies of culture and history, with nothing "real" beneath it? Are the epistemic norms of science guaranteed to lead us to a knowledge of this world, and if so why? Or are the epistemic norms also simply part of culture in the end, on a par with the metaphors of science? I worry about these questions, and now candor forces me to admit that—on the evidence we have—one could reasonably argue for either realism or nonrealism!

Suppose you go with Popper. With him, you believe that a real world exists "out there," independently of us. You are a metaphysical realist. You believe that we may never know the real world exactly, but "truth" is the correspondence of our ideas with this world, and the aim and method of science is to approach such truth, if only asymptotically. In your view, then, the epistemic norms truly do guarantee the approach to knowledge of reality as it is in itself. In the words of one of today's most eminent physicists, Steven Weinberg (1996, 14): "I have come to think that the laws of physics are real because my experience with the laws of physics does not seem to me to be very different in any fundamental way from my experience with rocks. For those who have not lived with the laws of physics, I can offer the obvious argument that the laws of physics as we know them work, and there is no other known way of looking at nature that works in anything like the same sense."

Surely, there is nothing to stop the Popperian from arguing that his or her philosophy applies equally well to the history of evolutionary thought. The importance of metaphor notwithstanding, an underlying reality still holds things up in the biological world just as it does in the physical world. We may never get to that reality, but it is there nonetheless. Either, most likely, one metaphor will triumph over all others and with time become more literally true of reality, or you conclude that different people can have different, but valid, perspectives on the same reality, just as two folk in different parts of Paris might have different perspectives on the Eiffel Tower. Although, continues the Popperian, let us not be too pessimistic. For all of its cultural elements, Darwinism is surely closer to reality than is creationism. If you think otherwise, you explain why the predictions of Parker and Sepkoski work so well. Is it just a miracle? And the same applies to the epistemic norms which lead to science. To continue with the words of our physicist, Weinberg: "There is also the related argument that although we have not yet had a chance to compare notes with the creatures on a distant planet, we can see that on Earth the laws of physics are understood in the same way by scientists of every nation, race, and—yes—gender" (14–15). Our study points precisely this way. Erasmus Darwin was condemned by exactly the criteria that Geoffrey Parker and Jack Sepkoski are praised.

But now swing over to the Kuhnians. For them, there is no reality, other than that seen through and created by the paradigm. I doubt Kuhn would be happy to be called an "idealist"—philosophers nowadays rarely are—but his realism is much softer than Popper's. There is nothing except as filtered through our perception or thought, and this in a sense means nothing outside our perception and thought. Hence, although Kuhn says little on this, his theory of truth is much more one of coherence than of correspondence. The aim is to get everything to hang together, for there is no external gauge against which to measure things. For Kuhn or a Kuhn-type philosopher, the norms have a more completely cultural status. Perhaps they are supercultural beliefs, of a kind that transcend the normal changes of culture. Or some such thing. After all, there is no real reason why all of culture has to conform to exactly the same pattern, or why some elements of culture should not be more long-lasting than other elements. Think of the Catholic Church. Some things change—the use of Latin, for instance. Other things remain unchanged, far longer and wider than our epistemic norms—the celibacy of the priesthood, for instance.

But surely again, you can readily apply this philosophy to the history of evolutionary thought. Ultimately you are no closer to some kind of absolute reality at the end than you were before you started. Your metaphors and theories may be more sophisticated, but if anything you are even more metaphorical and theoretical than when you started. The cultural layer between you and the world—if it existed—is thicker than it ever was. Are evolutionarily stable strategies any less an artifact of culture than the balance of nature? In this sense, evolution's history is no more progressive than is evolution itself, and even the vaunted epistemic values are no more than reflections of a particular age. It is true that things like prediction and unification are cherished right through our period, but absolutely no proof whatsoever has been offered to show that these values are not the creation of the Enlightenment, something tailor-made for an industrial secular society, as has characterized Western civilization for the past two or three centuries. It is not a question of saying that things are unreal or that they are false—dinosaurs are real and true, unicorns are unreal and false—but rather that reality, whatever it may be, simply does not make sense except in the context of an observer. The whole point about evolutionary theory is that the observer is involved throughout.

In this context—a point of which Kuhn himself makes much—remember the bitter disputes between Wilson and his critics, for example. If there were a real-world touchstone, one would not expect such disputes. But if all is a question of persuasion from within the position, as in politics, then such disagreement is expected. Only a partisan can believe that an independent world really exists out there which corresponds to Wilson's vision rather than Lewontin's, or conversely.

Nor is the threat of creationism so terrifying to a Kuhnian. The fact is that creationism is rightly rejected because it does not do as well epistemically as evolution. The Kuhnian is not denying the standards of science, just interpreting them differently. Of course, the subjectivist is more tolerant of multiple perspectives, and there is no doubt that many find acceptance of such diversity troubling—no less so in science than it is in areas like morality. But no less in science than in morality is diversity a reality, and to pretend otherwise is simply to perpetuate prejudice.

Everything in evolution's history confirms what the philosopher Hilary Putnam has said about his own Kuhnian-type theory of "internal realism": "'Truth', in an internalist view, is some sort of (idealized) rational acceptability—some sort of ideal coherence of our beliefs with each other and with our experiences *as those experiences are themselves represented in our belief system*—and not correspondence with mind-independent or discourse-independent 'states of affairs'. There is no God's Eye point of view that we can know or usefully imagine; there are only the various points of view of actual persons reflecting various interests and purposes that their descriptions and theories subserve" (Putnam 1981, 49–50).

Who is right: Dawkins with his functionalism or Gould with his transcendentalism? Lewontin with his tight antireductionism or Wilson with his daring expansionism? Parker with his Darwinism or Sepkoski with his Spencerianism? They are all right, at least as far as their theories work, and they are all wrong, at least as far as their theories do not work. Some things are better than others. We are not giving up on that. But there are no absolutes—certainly no absolutes against which we can declare one side right and the other side wrong. Even as we read now, some bright graduate student somewhere is probably about to make his or her name by resuscitating group selection.

Two positions: realism and nonrealism. And essentially we seem to be no further ahead than when we started! Apparently, our history does not decide between the two. I would go further: our history does not even begin to decide between the two. A naturalistic approach as we have taken—going out and looking at the evidence, as a scientist might look at the evidence—an approach which was accepted if not urged on us by partisans from both sides in the current controversy over the nature of science, is not going to work. I am certainly not now saying that the debate over realism and nonrealism (idealism) is unimportant or that partisans cannot offer good arguments for their respective sides. Without starting a whole new line of inquiry, my inclination with respect to this question would be to turn to more traditional philosophical tools and forms of argument—not really such a radical suggestion if it is indeed a philosophical problem which is at stake. One might fruitfully start by asking about matters of meaning. What would it mean to talk of an entity which exists even if it is or were unobserved by humankind? Can one draw significant analogies between unobserved (and probably unobservable) entities like electrons and unobserved nature in general? And so forth. (See Ruse 1986 and Klee 1997 for more thoughts on these matters.)

Perhaps by taking such a philosophical approach, some answers will be forthcoming on the realism/nonrealism issue. Perhaps not. At least it offers some avenues which our dead-end approach does not. These are paths to be explored in another place at another time. For us, the time has come to stop worrying, and to start taking our failure as a reason to move on: perhaps indeed to take it as a cause for celebration rather than as a cause for despair. For our history is certainly relevant to something. A lot of people (not just me!) have put a lot of time and effort into ferreting out the facts about the history of evolutionary ideas and how and in what directions they have changed over the years. We know much about the theoretical concepts of evolution; and in recent years, thanks in no small part to constructivists and their allies, we have learned much about the social structure of evolutionism and of the people behind it. So the nonrelevance of our history—at least, the nondecisive nature of our history—to the traditional philosophical debate suggests to me that the bitter divide between the scientists (objectivists) and their critics (constructivists or subjectivists) is operating at a different level.

The debate in the first place is about the integrity of professional science as something with disinterested standards, and the difference between science at this level and popular science or pseudo-science or any other flight of fancy (including religion and philosophy and much else) that you might care to consider. The debate is about good-quality science and bad-quality science, or about science and stuff which should not be considered science at all, no matter what its partisans claim. Here the Popperian wins; but there is no reason why the Kuhnian should not accept the victory and claim that this is what he or she meant all along! Both the realist and the nonrealist can make the distinctions between good and bad (or non-) science, and for the same reasons. The nonrealist or internal realist can draw no less firm a line in the sand than the metaphysical realist between (say) the work of Geoffrey Parker and Erasmus Darwin, let alone mesmerism—and he or she does. Within the system, the Kuhnian can talk just as much about objectivity as can the Popperian, and about good, professional, mature science as opposed to the failed contenders or those which are not in contention at all (Ruse 1986).

I am not saying that the nonrealist is always as careful on this point as might be warranted. Indeed, with the recent fascination with such topics as phrenology, one might well say that the waters have been muddied, significantly. But distinctions can be drawn. That belief, in fact, is as much a presupposition and starting point of *The Structure of Scientific Revolutions* as it is of *The Logic of Scientific Discovery*. (See Kuhn 1977 for an explicit statement about the importance of epistemic standards.)

Then, in the second place, the debate is about the connection between culture and science. The Kuhnian critics were right in showing that, with respect to culture, science is not that different from the rest of human experience. In its way, science is no less cultural than the other products of the human mind. And when you think about it, the critics are surely right in saying that only people who have been indoctrinated by their culture to think that they are above that culture could possibly hold such a very odd belief. But here again, victory need not trouble the other side, the Popperians. Since culture adds to and aids what they find valuable, they can give way to their subjectivist critics. Anything which makes for fertile and ongoing science is as much the presupposition and starting point of *The Logic of Scientific Discovery* as it is of *The Structure of Scientific Revolutions*.

My point therefore is that two different debates have been confused. There is the old philosophical debate about realism/nonrealism. Nothing inferred from the history of science can speak directly to this. Or, if you protest that I am overstating the case, nothing inferred from our history of science can be decisive on this. Then there is the new debate about standards and culture. Is science something special on its own, separate from other disciplines and from its pretenders? And is this difference in part (or whole) because science is subject to certain demanding standards, against which its successes are measured? Are these standards such that we would speak of science as "objective," meaning beyond the individual's whims? Perhaps, if you are adding some pragmatic dimension, such as enabling us to put men on the moon or to cure childhood cancers or to vaporize thousands of Japanese. And is science beyond culture? Or must we still speak of science as "subjective," meaning that, because of its cultural impregnation, if it pleases you I have no grounds to criticize? The history of science does speak to this debate. And its answer, obtained within the context of our story, is as follows: It is true that science is special, and this is because of its standards; the critics were wrong in arguing otherwise. But it is also true that science is not special, and this is because of its culture; the defenders were wrong in arguing otherwise.

So, in the end, one can (whether a metaphysical realist or not) talk of objectivity and subjectivity, reality or the nonreal—for this is the reality of "reality versus illusion: Macbeth's dagger right there in the room or not," not the reality of "the noise a tree makes as it falls in the forest, when there is no one around to hear it." Good-quality science tells us about this former kind of reality. Poor-quality theories or discourses—pseudo- or quasi-sciences—do not. This is true whether you think there is ultimately a kind of human-independent reality or not. Within the system, the Kuhnian no less than the Popperian can distinguish between the real and the fake or chimerical. It is just that ultimately, talking *about* rather than *within* the system, reality for the Kuhnian is coherence rather than correspondence.

EPILOGUE

Terms of Engagement

Is the history of evolutionary theory recounted in these chapters typical of the rest of science? Are my claims about metaphor and objectivity applicable to the social sciences? To the physical sciences? The limits of an empirical, naturalistic approach are that you cannot generalize beyond the data examined, just because a generalization seems plausible or desirable. My suspicion is that probably my story does have wider relevance. Certainly, I am much aware of the significance of metaphor in other areas of science. But I will conclude with the modest claim that in the key area of evolutionary biology we can resolve the debate over the nature of science.

Or perhaps this claim is not so modest. Before the argument continues, others should do for the history of their subject what I have done for mine. Do not simply throw at us disgusting stories about the personal lives of the great men of science or of the vile or outlandish values that they embraced. Move the debate forward now and show that I am wrong in claiming that cultural values do get pushed out, that epistemic norms are important and stand up through time and space, that popularizers, however much they are respected, are nevertheless regarded as popularizers, that culture persists even if the values do not, and more. Do these things and then we can argue again.

ILLUSTRATION CREDITS

REFERENCES

GLOSSARY

INDEX

Photograph Credits

Karl Popper, photograph by permission of Open Court Publishing, Inc.

Thomas Kuhn, photograph by permission of MIT Museum

Erasmus Darwin, portrait by J. Wright, photograph by permission of the National Portrait Gallery

Charles Darwin, watercolor by George Richmond, photograph by permission of English Heritage Photographic Library

Julian Huxley, photograph by permission of Woodson Research Center, Fondren Library, Rice University

Theodosius Dobzhansky, photograph from *Dobzhansky's Genetics of Natural Populations* IXLIII, ed. R. C. Lewontin, J. A. Moore, W. B. Provine, and B. Wallace (New York: Columbia University Press, 1981), by permission of the publisher

Richard Dawkins, photograph by Lisa Lloyd, by permission

Stephen Jay Gould, photograph by Jon Chase, by permission of the Harvard News Office

Richard Lewontin, photograph by Rick Stafford, by permission of the Harvard News Office

Edward O. Wilson, photograph by Jon Chase, by permission of the Harvard News Office

Geoffrey Parker, photograph by permission of Geoffrey Parker

Jack Sepkoski, photograph by permission of Jack Sepkoski

References

Achinstein, P., and Barker S. F., eds. 1969. *The Legacy of Logical Positivism: Studies in the Philosophy of Science*. Baltimore: Johns Hopkins University Press.

Allen, G. E. 1978a. *Life Science in the Twentieth Century*. Cambridge: Cambridge University Press.

——. 1978b. *Thomas Hunt Morgan: The Man and His Science*. Princeton, N.J.: Princeton University Press.

Alvarez, L. W., W. Alvarez, F. Asaro, and H. V. Michel. 1980. Extraterrestrial cause for the Cretaceous-Tertiary extinction. *Science* 208: 1095–108.

Ayala, F. J. 1974. The concept of biological progress. In *Studies in the Philosophy of Biology*, ed. F. J. Ayala and T. Dobzhansky. London: Macmillan.

Ayala, F. J., M. L. Tracey, L. G. Barr, J. F. McDonald, and S. Perez-Salas. 1974. Genetic variation in natural populations of five Drosophila species and the hypothesis of the selective neutrality of protein polymorphisms. *Genetics* 77: 343–384.

Babbage, C. [1838] 1967. *The Ninth Bridgewater Treatise: A Fragment*. 2nd ed. London: Frank Cass.

Baker, J. R. 1976. Julian Sorell Huxley. *Biographical Memoirs of Fellows of the Royal Society* 22: 207–38.

Baker, R. R., and G. A. Parker. 1979. Unprofitable prey. *New Scientist* 1185: 898.

Barbour, I. 1988. Ways of relating science and theology. In *Physics, Philosophy, and Theology: A Common Quest for Understanding*, ed. R. J. Russell, W. R. Stoeger, and G. V. Coyne. Vatican City: Vatican Observatory.

Barrett, P., P. J. Gautrey, S. Herbert, D. Kohn, and S. Smith, eds. 1987. *Charles Darwin's Notebooks: 1836–1844*. Ithaca: Cornell University Press.

Bates, H. W. 1862. Contributions to an insect fauna of the Amazon valley. *Transactions of the Linnaean Society of London* 23: 495–566.

Bateson, B. 1928. *William Bateson, F.R.S., Naturalist: His Essays and Addresses together with a Short Account of His Life*. Cambridge: Cambridge University Press.

Beer, G. 1983. *Darwin's Plots: Evolutionary Narrative in Darwin, George Eliot, and Nineteenth Century Fiction*. London: Routledge and Kegan Paul.

Bergson, H. 1911. *Creative Evolution*. London: Macmillan.

Bloor, D. 1976. *Knowledge and Social Imagery.* London: Routledge and Kegan Paul.

Box, J. F. 1978. *R. A. Fisher: The Life of a Scientist.* New York: Wiley.

Bowler, P. 1984. *Evolution: The History of an Idea.* Berkeley: University of California Press.

———. 1988. *The Non-Darwinian Revolution: Reinterpreting a Historical Myth.* Baltimore, Md.: Johns Hopkins University Press.

———. 1996. *Life's Splendid History.* Chicago: University of Chicago Press.

Broad, C. D. 1949. Review of Julian S. Huxley's *Evolutionary Ethics.* In *Readings in Philosophical Analysis,* ed. H. Feigel and W. Sellars. New York: Apple-Century-Crofts (org. pub. *Mind* 53 [1944]).

Buchwald, J. 1989. *The Rise of the Wave Theory of Light: Optical Theory and Experiment in the Early Nineteenth Century.* Chicago: Chicago University Press.

Burchfield, J. D. 1974. Darwin and the dilemma of geological time. *Isis* 65: 300–331.

———. 1975. *Lord Kelvin and the Age of the Earth.* New York: Science History Publications.

Burkhardt, R. W. 1977. *The Spirit of System: Lamarck and Evolutionary Biology.* Cambridge, Mass.: Harvard University Press.

Cain, A. J., and P. M. Sheppard. 1950. Selection in the polymorphic land snail Cepaea nemoralis. *Heredity* 4: 275–94.

———. 1952. The effects of natural selection on body colour in the land snail Cepaea nemoralis. *Heredity* 6: 217–31.

———. 1954. Natural selection in Cepaea. *Genetics* 39: 89–116.

Cain, J. A. 1992. Common problems and cooperative solutions: organizational activity in evolutionary studies 1936–1947. *Isis* 84: 1–25.

———. 1994. Ernst Mayr as community architect: Launching the Society for the Study of Evolution and the journal *Evolution. Biology and Philosophy* 9 (3): 387–428.

Callebaut, W. 1994. *Taking the Naturalistic Turn.* Chicago: University of Chicago Press.

Campbell, D. T. 1974. Evolutionary epistemology. In *The Philosophy of Karl Popper,* ed. P. A. Schilpp. LaSalle, Ill.: Open Court.

Canning, G., H. Frere, and G. Ellis. 1798. The loves of the triangles. *Anti-Jacobin*: April 16, April 23, and May 17.

Cannon, W. B. 1931. *The Wisdom of the Body.* Cambridge, Mass.: Harvard University Press.

Cannon, W. F. 1961. The impact of uniformitarianism: Two letters from John Herschel to Charles Lyell, 1836–1837. *Proceedings of the American Philosophical Society* 105: 301–14.

Clarke, B. 1974. Causes of genetic variation. Review of R. C. Lewontin, *The Genetic Basis of Evolutionary Change. Science* 186: 584–85.

Clifford, W. K. 1879. *Lectures and Essays,* ed. L. Stephen and F. Pollack. London: Macmillan.

Clutton-Brock, T. H., F. E. Guinness, and S. D. Albon. 1982. *Red Deer: Behaviour and Ecology of the Two Sexes.* Chicago: University of Chicago Press.

Clutton-Brock, T. H., and G. A. Parker. 1995a. Punishment in animal societies. *Nature* 373: 209–16.

———. 1995b. Sexual coercion in animal societies. *Animal Behaviour* 49 (5): 1345–65.

Cooter, R. 1984. *The Cultural Meaning of Popular Science: Phrenology and the Organization of Consent in Nineteenth-Century Britain.* Cambridge: Cambridge University Press.

Crook, P. 1994. *Darwinism: War and History.* Cambridge: Cambridge University Press.

Cuvier, G. 1810. *Rapport historique sur les progrès des sciences naturelles.* Paris.

———. 1813. *Essay on the Theory of the Earth,* trans. R. Kerr. Edinburgh: W. Blackwood.

Darnton, R. 1968. *Mesmerism and the End of the Enlightenment in France.* Cambridge, Mass.: Harvard University Press.

Darwin, C. 1859. *On the Origin of Species.* London: John Murray.

———. 1868. *The Variation of Animals and Plants under Domestication.* 2 vols. London: John Murray.

———. 1871. *The Descent of Man.* 2 vols. London: John Murray.

———. April 27, 1871. Letter to the Editor. *Nature* 3:502–503.

———. 1872. *On the Origin of Species.* 6th ed.

———. 1959. *The Origin of Species by Charles Darwin: A Variorum Text,* ed. M. Peckham. Philadelphia: University of Pennsylvania Press.

———. [1873] 1977. On the males and complemental males of certain Cirripedes, and on rudimentary structures. In *The Collected Papers of Charles Darwin,* ed. P. H. Barrett. Chicago: University of Chicago Press.

Darwin, C., and A. R. Wallace. 1958. *Evolution by Natural Selection.* Foreword by G. de Beer. Cambridge: Cambridge University Press.

Darwin, E. 1794–1796. *Zoonomia; or, The Laws of Organic Life.* London: J. Johnson.

———. 1801. *Zoonomia; or, The Laws of Organic Life.* 3rd ed. London: J. Johnson.

———. 1803. *The Temple of Nature.* London: J. Johnson.

Davis, M., P. Hut, and R. A. Muller. 1984. Extinction of species by periodic comet showers. *Nature* 308: 715–17.

Dawkins, R. 1976. *The Selfish Gene.* Oxford: Oxford University Press.

———. 1986. *The Blind Watchmaker.* New York: Norton.

———. 1989. *The Selfish Gene.* 2nd ed. Oxford: Oxford University Press.

———. 1995. *A River Out of Eden.* New York: Basic Books.

———. 1996. *Climbing Mount Improbable.* New York: Norton.

Derrida, J. 1970. Structure, sign, and play in the discourse of the human sciences. In *The Languages of Criticism and the Sciences of Man: The Structuralist Controversy,* ed. R. Macksey and E. Donato. Baltimore, Md.: Johns Hopkins University Press.

Desmond, A. 1997. *Huxley, Evolution's High Priest.* London: Michael Joseph.

Desmond, A., and J. Moore. 1992. *Darwin: The Life of a Tormented Evolutionist.* New York: Warner.

Dobzhansky, T. 1937. *Genetics and the Origin of Species.* New York: Columbia University Press.

———. 1956. *The Biological Basis of Human Freedom.* New York: Columbia University Press.

———. 1962. *Mankind Evolving.* New Haven: Yale University Press.

———. 1967. *The Biology of Ultimate Concern.* New York: New American Library, Inc.

———. 1970. *Genetics of the Evolutionary Process.* New York: Columbia University Press.

———. [1943] 1981. Temporal changes in the composition of populations of Drosophila pseudoobscura in different environments. *Genetics* 28: 162–86. In *Dobzhansky's Genetics of Natural Populations I–XLIII*, ed. R. C. Lewontin, J. A. Moore, W. B. Provine, and B. Wallace. New York: Columbia University Press.

Dobzhansky, T., F. J. Ayala, G. L. Stebbins, and J. W. Valentine. 1977. *Evolution.* San Francisco: Freeman.

Dobzhansky, T., and H. Levene. 1955. Developmental homeostasis in natural populations of Drosophila pseudoobscura. *Genetics* 40: 797–808.

Dobzhansky, T., and B. Wallace. 1953. The genetics of homeostasis in Drosophila. *Proceedings of the National Academy of Sciences* 39: 162–71.

———. 1959. *Radiation, Genes and Man.* New York: Henry Holt and Company.

Duhem, P. [1906] 1954. *The Aim and Structure of Physical Theory.* Trans. Philip P. Wiener. Princeton, N.J.: Princeton University Press.

Egerton, F. N. 1973. Changing concepts of the balance of nature. *Quarterly Review of Biology* 48 (June): 322–50.

Eldredge, N. 1971. The allopatric model and phylogeny in paleozoic invertebrates. *Evolution* 25: 156–67.

Eldredge, N., and S. J. Gould. 1972. Punctuated equilibria: an alternative to phyletic gradualism. In *Models in Paleobiology*, ed. T. J. M. Schopf. San Francisco, Calif.: Freeman, Cooper.

Ellegard, A. 1958. *Darwin and the General Reader.* Goteborg: Goteborgs Universitets Arsskrift.

Ferguson, H. 1990. *The Science of Pleasure: Cosmos and Psyche in the Bourgeois World View.* London: Routledge.

Fish, S. 1996. Professor Sokal's bad joke. *New York Times*, May 21: 23.

Fisher, R. A. 1930. *The Genetical Theory of Natural Selection.* Oxford: Clarendon Press.

Fodor, J. 1996. Peacocking. *London Review of Books*, April 18: 19–20.

Foucault, M. 1970. *The Order of Things: An Archaeology of the Human Sciences.* New York: Pantheon.

Franklin, B., A. Lavoisier, and others. [1784] 1996. Rapport des Commissaires chargés par LE ROI de l'Examen du Magnétisme animal. *Skeptic* 4 (3): 68–83.

Geison, G. L. 1969. Darwin and heredity: the evolution of his hypothesis of pangenesis. *Bulletin of the History of Medicine* 24: 375–411.

———. 1995. *The Private Life of Louis Pasteur.* Princeton: Princeton University Press.

Gillispie, C. 1950. *Genesis and Geology.* Cambridge, Mass.: Harvard University Press.

Glen, W. 1994. *The Mass-Extinction Debates: How Science Works in a Crisis.* Stanford: Stanford University Press.

Godfray, H. C. J., and G. A. Parker. 1991. Clutch size, fecundity, and parent-offspring conflict. In *The Evolution of Reproductive Strategies,* ed. P. H. Harvey, L. Partridge, and T. R. E. Southwood, pp. 67–79. London: Philosophical Transactions of the Royal Society.

———. 1992. Sibling competition, parent-offspring conflict and clutch size. *Animal Behaviour* 43 (3): 473–90.

Gould, S. J. 1966. Allometry and size in ontogeny and phylogeny. *Biological Reviews of the Cambridge Philosophical Society* 41: 587–640.

———. 1969. An evolutionary microcosm: Pleistocene and Recent history of the land snail *P. (Poecilozonites)* in Bermuda. *Bulletin of the Museum of Comparative Zoology* 138: 407–532.

———. 1977a. *Ever Since Darwin.* New York: Norton.

———. 1977b. *Ontogeny and Phylogeny.* Cambridge, Mass.: Belknap Press of Harvard University Press.

———. 1979. Episodic change versus gradualist dogma. *Science and Nature* 2: 5–12.

———. 1980a. Is a new and general theory of evolution emerging? *Paleobiology* 6: 119–30.

———. 1980b. Sociobiology and the theory of natural selection. In *Sociobiology: Beyond Nature/Nurture?* ed. G. Barlow and J. Silverberg. Boulder, Col.: Westview.

———. 1981. *The Mismeasure of Man.* New York: Norton.

———. 1982a. Darwinism and the expansion of evolutionary theory. *Science* 216: 380–7.

———. 1982b. The meaning of punctuated equilibrium and its role in validating a hierarchical approach to macroevolution. In *Perspectives on Evolution,* ed. R. Milkman. Sunderland, Mass.: Sinauer.

———. 1983. Irrelevance, submission, partnership: the changing role of paleontology in Darwin's three centennials and a modest proposal for macroevolution. *Evolution from Molecules to Men,* ed. D. S. Bendall. Cambridge: Cambridge University Press.

———. 1984. Morphological channeling by structural constraint: convergence in styles of dwarfing and giantism in Cerion, with a description of two new fossil species and a report on the discovery of the largest Cerion. *Paleobiology* 10: 172–94.

———. 1989. *Wonderful Life: The Burgess Shale and the Nature of History*. New York: W. W. Norton Co.

———. 1996. *Full House: The Spread of Excellence from Plato to Darwin*. New York: Paragon.

———. 1997. Darwinian fundamentalism. *New York Review of Books* 44 (10): 34–37.

Gould, S. J., and C. B. Calloway. 1980. Clams and brachiopods—ships that pass in the night. *Paleobiology* 6: 383–96.

Gould, S. J., and N. Eldredge. 1977. Punctuated equilibria: the tempo and mode of evolution reconsidered. *Paleobiology* 3: 115–51.

Gould, S. J., and R. C. Lewontin. 1979. The spandrels of San Marco and the Panglossian paradigm: a critique of the adaptationist program. *Proceedings of the Royal Society of London, Series B: Biological Sciences* 205: 581–98.

Gould, S. J., D. M. Raup, J. J. Sepkoski Jr, T. J. M. Schopf, and D. M. Simberloff. 1977. The shape of evolution: a comparison of real and random clades. *Paleobiology* 3: 23–40.

Gould, S. J., and E. S. Vrba. 1982. Exaptation—a missing term in the science of form. *Paleobiology* 8: 4–15.

Gray, A. 1876. *Darwiniana*. New York: D. Appleton.

Greene, J. C., and M. Ruse. 1996. On the nature of the evolutionary process: the correspondence between Theodosius Dobzhansky and John C. Greene. *Biology and Philosophy* 11: 445–91.

Gross, A. 1990. *The Rhetoric of Science*. Cambridge, Mass.: Harvard University Press.

Gross, P. R., and N. Levitt. 1994. *Higher Superstition: The Academic Left and Its Quarrels with Science*. Baltimore, Md.: Johns Hopkins University Press.

Haeckel, E. 1866. *Generelle Morphologie der Organismen*. Berlin: Reimer.

———. 1868. *The History of Creation*. London: Kegan Paul, Trench.

———. 1896. *The Evolution of Man*. New York: Appleton.

Hallam, A. 1984. The causes of mass extinctions. *Nature* 308: 686–87.

Hamilton, W. D. 1964a. The genetical evolution of social behaviour I. *Journal of Theoretical Biology* 7: 1–16.

———. 1964b. The genetical evolution of social behaviour II. *Journal of Theoretical Biology* 7: 17–32.

———. 1996. *Narrow Roads of Gene Land: The Collected Papers of W. D. Hamilton*. New York: W. H. Freeman/Spektrum.

Hankin, C. C., ed. 1858. *Life of Mary Anne SchimmelPenninck*. 2 vols. London: Longman.

Henderson, L. J. 1913. *The Fitness of the Environment*. New York: Macmillan.

———. 1917. *The Order of Nature*. Cambridge, Mass.: Harvard University Press.

Hesse, M. 1966. *Models and Analogies in Science.* Notre Dame: University of Notre Dame Press.

Hodge, M. J. S. 1992. Biology and philosophy (including ideology): a study of Fisher and Wright. In *The Founders of Evolutionary Genetics,* ed. S. Sarkar. Dordrecht: Kluwer Academic Publishers.

Hoffman, A. 1985. Patterns of family extinction depend on definition and geological timescale. *Nature* 315: 659–62.

———. 1986. Reply. *Nature* 321: 535–36.

———. 1989. *Arguments on Evolution: A Paleontologist's Perspective.* New York: Oxford University Press.

Hölldobler, B., and E. O. Wilson. 1990. *The Ants.* Cambridge, Mass.: Harvard University Press.

Howells, W. W. 1960. The distribution of man. *Scientific American* 203 (3): 112–127.

Hubby, J. L., and R. C. Lewontin. 1966. A molecular approach to the study of genic heterozygosity in natural populations. I. The number of alleles at different loci in Drosophila pseudoobscura. *Genetics* 54: 577–94.

Hull, D. 1973. *Darwin and His Critics.* Cambridge, Mass.: Harvard University Press.

Hume, D. [1779] 1947. *Dialogues Concerning Natural Religion,* ed. N. K. Smith. Indianapolis, Ind.: Bobbs-Merrill Co.

Huxley, J. S. 1932. *Problems of Relative Growth* . London: Methuen.

———. 1934. *Bird-Watching and Bird Behaviour.* London: Chatto and Windus.

———. 1936. Natural selection and evolutionary progress. *British Association for the Advancement of Science, Report of the Annual Meeting, 1936, Blackpool, September 9–16*: 81–100.

———. 1942. *Evolution: The Modern Synthesis.* London: Allen and Unwin.

———. 1943. *TVA: Adventure in Planning.* London: Scientific Book Club.

———. 1948. *UNESCO: Its Purpose and Its Philosophy.* Washington, D.C.: Public Affairs Press.

———. 1970. *Memories.* London: Allen and Unwin.

———. 1973. *Memories II.* London: Allen and Unwin.

Huxley, L. 1900. *The Life and Letters of Thomas Henry Huxley.* 2 vols. London: Macmillan.

Huxley, T. H. 1869. Anniversary address of the president. *Quarterly Journal of the Geological Society of London* 25: xxviii–liii.

Hyman, S. E. 1962. *The Tangled Bank: Darwin, Marx, Frazer and Freud as Imaginative Writers.* New York: Atheneum.

Johanson, D., and M. Edey. 1981. *Lucy: The Beginnings of Humankind.* New York: Simon and Schuster.

Johnson, P. E. 1995. *Reason in the Balance.* Downers Grove, Ill.: InterVarsity Press.

Joravsky, D. 1970. *The Lysenko Affair.* Cambridge, Mass.: Harvard University Press.

Kevles, D. J. 1992. Huxley and the popularization of science. In *Julian Huxley: Biologist and Statesman of Science*, ed. C. K. Waters and A. Van Helden. Houston: Rice University Press.

Kimmelman, B. A. 1987. A Progressive Era Discipline: Genetics at American Agricultural Colleges and Experimental Stations, 1900–1920. PhD Dissertation, Pennsylvania.

King-Hele, D. 1963. *Erasmus Darwin: Grandfather of Charles Darwin*. New York: Scribners.

———, ed. 1981. *The Letters of Erasmus Darwin*. Cambridge: Cambridge University Press.

Kitchell, J. A., and G. Estabrook. 1986. Reply to Hoffman. *Nature* 321: 534–35.

Kitcher, P. 1985. *Vaulting Ambition*. Cambridge, Mass.: M.I.T. Press.

Klee, R. 1997. *Introduction to the Philosophy of Science: Cutting Nature at Its Seams*. New York: Oxford University Press.

Kohler, R. 1996. Book review of G. Geison, *The Private Life of Louis Pasteur*. *Isis* 87: 331–34.

Kropotkin, P. [1902] 1955. *Mutual Aid*, ed. A. Montague. Boston, Mass.: Extending Horizons Books.

Kuhn, T. 1957. *The Copernican Revolution*. Cambridge, Mass.: Harvard University Press.

———. 1962. *The Structure of Scientific Revolutions*. Chicago: University of Chicago Press.

———. 1977. Objectivity, value, judgement, and theory choice. In *The Essential Tension: Selected Studies in Scientific Tradition and Change*, ed. T. Kuhn. Chicago: University of Chicago Press.

———. 1993. Metaphor in science. In *Metaphor and Thought*, 2nd ed., ed. A. Ortony, pp. 533–42. Cambridge: Cambridge University Press.

Labandeira, C. C., and J. J. Sepkoski Jr. 1993. Insect diversity in the fossil record. *Science* 261: 310–315.

Lakatos, I. 1970. Falsification and the methodology of scientific research programmes. In *Criticism and the Growth of Knowledge*, ed. I. Lakatos and A. Musgrave. Cambridge: Cambridge University Press.

Lakoff, G., and M. Johnson. 1980. *Metaphors We Live By*. Chicago: University of Chicago Press.

Latour, B., and S. Woolgar. 1979. *Laboratory Life: The Construction of Scientific Facts*. Beverly Hills, Calif.: Sage.

Laudan, L. 1981. *Science and Hypothesis*. Dordrecht: D. Reidel.

Laudan, R. 1987. *From Mineralogy to Geology: The Foundations of a Science, 1650–1830*. Chicago: University of Chicago Press.

Lerner, I. M. 1954. *Genetic Homeostasis*. New York: John Wiley.

Levins, R., and R. C. Lewontin. 1985. *The Dialectical Biologist*. Cambridge, Mass.: Harvard University Press.

Lewontin, R. C. 1957. The adaptation of populations to varying environments. *Cold Spring Harbor Symposia on Quantitative Biology* 22: 395–408.

―――. 1974. *The Genetic Basis of Evolutionary Change.* New York: Columbia University Press.

―――. 1977. Sociobiology—a caricature of Darwinism. In *PSA 1976*, ed. F. Suppe and P. Asquith. East Lansing, Mich.: Philosophy of Science Association.

―――. 1982. *Human Diversity.* New York: Scientific American Library.

―――. 1985. Population genetics. *Annual Review of Genetics* 19: 81–102.

―――. 1991a. Twenty-five years ago in genetics—electrophoresis in the development of evolutionary genetics—milestone or millstone? *Genetics* 128: 657–62.

―――. 1991b. *Biology as Ideology: The Doctrine of DNA.* Toronto: Anansi.

―――. 1997. Billions and billions of demons. *New York Review of Books* 44 (1): 28–32.

Lewontin, R. C., and J. L. Hubby. 1966. A molecular approach to the study of genic heterozygosity in natural populations. II. Amount of variation and degree of heterozygosity in natural populations of Drosophila pseudoobscura. *Genetics* 54: 595–609.

Lewontin, R. C., S. Rose, and L. J. Kamin. 1984. *Not in Our Genes: Biology, Ideology and Human Nature.* New York: Pantheon.

Longino, H. 1990. *Science as Social Knowledge.* Princeton, N.J.: Princeton University Press.

Lumsden, C. J., and E. O. Wilson. 1981. *Genes, Mind, and Culture.* Cambridge, Mass.: Harvard University Press.

―――. 1983. *Promethean Fire: Reflections on the Origin of Mind.* Cambridge, Mass.: Harvard University Press.

MacArthur, R. H., and E. O. Wilson. 1963. Equilibrium theory of island zoogeography. *Evolution* 17: 373–87.

―――. 1967. *The Theory of Island Biogeography.* Princeton N.J.: Princeton University Press.

Maddox, J. 1985. Periodic extinctions undermined. *Nature* 315: 627.

Marshall, L. G., S. D. Webb, J. J. Sepkoski Jr., and D. M. Raup. 1982. Mammalian evolution and the great American interchange. *Science* 215: 1351–57.

Maynard Smith, J. 1978. The evolution of behavior. *Scientific American* 239 (3): 176–193.

―――. 1982. *Evolution and the Theory of Games.* Cambridge: Cambridge University Press.

―――. 1995. Genes, memes, and minds. *New York Review of Books* 42 (19): 46–48.

Mayr, E. 1942. *Systematics and the Origin of Species.* New York: Columbia University Press.

―――. 1959. Where are we? *Cold Spring Harbor Symposia on Quantitative Biology* 24: 1–14.

————. 1963. *Animal Species and Evolution.* Cambridge, Mass.: Harvard University Press.

————. 1982. *The Growth of Biological Thought: Diversity, Evolution and Inheritance.* Cambridge, Mass.: Belknap Press of Harvard University Press.

————. 1988. *Towards a New Philosophy of Biology: Observations of an Evolutionist.* Cambridge, Mass.: Belknap Press of Harvard University Press.

McMullin, E. 1983. Values in Science. In *PSA 1982,* ed. P. D. Asquith and T. Nickles, pp. 3–28. East Lansing, Mich.: Philosophy of Science Association.

McNeill, M. 1987. *Under the Banner of Science: Erasmus Darwin and His Age.* Manchester: Manchester University Press.

Medawar, P. [1961] 1967. Review of *The Phenomenon of Man.* In *The Art of the Soluble,* ed. P. Medawar. London: Methuen and Co. Ltd.

Milne-Edwards, H. 1827. Organisation. In *Dictionnaire Classique d'Histoire Naturelle.*

————. 1834. *Elements de zoologie: leçons sur l'anatomie, la physiologie, la classification des moeurs des animaux.* Paris: Crochard.

Mitman, G. 1992. *The State of Nature: Ecology, Community, and American Social Thought, 1900–1950.* Chicago: University of Chicago Press.

Muller, H. J. 1949. The Darwinian and modern conceptions of natural selection. *Proceedings of the American Philosophical Society* 93: 459–70.

Muller, H. J., and R. Falk. 1961. Are induced mutations in Drosophila overdominant? I. Experimental design. *Genetics* 46: 727–35.

Myers, G. 1990. *Writing Biology: Texts in the Social Construction of Scientific Knowledge.* Madison, Wisc.: University of Wisconsin Press.

Numbers, R. 1992. *The Creationists.* New York: A. A. Knopf.

Nyhart, L. K. 1995. *Biology Takes Form: Animal Morphology and the German Universities.* Chicago: University of Chicago Press.

Osborn, H. F. 1934. Aristogenesis: the creative principle in the origin of species. *American Naturalist* 68: 193–235.

Outram, D. 1984. *Georges Cuvier: Vocation, Science and Authority in Post-Revolutionary France.* Manchester: Manchester University Press.

Parker, G. A. 1969. The reproductive behaviour and the nature of sexual selection in Scatophaga stercovaria L. (Diptera: Scatophagidae). III. Apparent intersex individuals and their evolutionary cost to normal searching males. *Transactions of the Royal Entomological Society, London*: 305–23.

————. 1970a. The reproductive behaviour and the nature of sexual selection in Scatophaga stercovaria L. (Diptera: Scatophagidae). I. Diurnal and seasonal changes in population density around the site of mating and oviposition. *Journal of Animal Ecology* 39: 185–204.

————. 1970b. The reproductive behaviour and the nature of sexual selection in Scatophaga stercovaria L. (Diptera: Scatophagidae). II. The fertilization rate and the spatial and temporal relationships of each sex around the site of mating and oviposition. *Journal of Animal Ecology* 39: 205–28.

———. 1970c. Sperm competition and its evolutionary effect on copula duration in the fly Scatophaga stercovaria. *Journal of Insect Physiology* 16: 1301–28.

———. 1970d. The reproductive behaviour and the nature of sexual selection in Scatophaga stercovaria L. (Diptera: Scatophagidae). IV. Epigamic recognition and competition between males for the possession of females. *Behaviour* 37: 113–39.

———. 1970e. The reproductive behaviour and the nature of sexual selection in Scatophaga stercovaria L. (Diptera: Scatophagidae). V. The female's behaviour at the oviposition site. *Behaviour* 37: 140–168.

———. 1970f. The reproductive behaviour and the nature of sexual selection in Scatophaga stercovaria L. (Diptera: Scatophagidae). VIII. The origin and evolution of the passive phase. *Evolution* 24: 744–88.

———. 1970g. Sperm competition and its evolutionary consequences in the insects. *Biological Reviews* 45: 525–67.

———. 1974a. The reproductive behaviour and the nature of sexual selection in Scatophaga stercovaria L. (Diptera: Scatophagidae). IX. Spatial distribution of fertilization rates and evolution of male search strategy within the reproductive area. *Evolution* 28: 93–108.

———. 1974b. Assessment stategy and the evolution of fighting behaviour. *Journal of Theoretical Biology* 47: 223–44.

———. 1978a. Selfish genes, evolutionary games, and adaptiveness of behaviour. *Nature* 274 (5674): 849–55.

———. 1978b. Searching for mates. Review article in *Behavioural Ecology: An Evolutionary Approach,* ed. J. R. Krebs and N. B. Davies. Oxford: Blackwell.

———. 1985. Models of parent-offspring conflict. 5. Effects of the behaviour of the two parents. *Animal Behaviour* 33 (May): 513–33.

Parker, G. A., and M. R. MacNair. 1978. Modes of parent-offspring conflict. 1. Monogamy. *Animal Behaviour* 26 (Feb.): 97–110.

———. 1979. Models of parent-offspring conflict. 4. Suppression—evolutionary retaliation by the parent. *Animal Behaviour* 27 (Nov.): 1210–35.

Parker, G. A., and J. Maynard Smith. 1990. Optimality theory in evolutionary biology. *Nature* 348: 27–33.

Parker, G. A., and R. G. Pearson. 1976. Possible origin and adaptive significance of mounting behaviour shown by some female mammals in estrus. *Journal of Natural History* 10 (3): 241–45.

Parker, G. A., and R. A. Stuart. 1976. Animal behaviour as a strategy optimizer—evolution of resource assessment strategies and optimal emigration thresholds. *American Naturalist* 110 (976): 1055–76.

Patterson, C., and A. B. Smith. 1987. Is the periodicity of extinctions a taxonomic artifact? *Nature* 330: 248–51.

Pearson, K. 1892. *The Grammar of Science.* London: Walter Scott.

Perutz, M. 1995. The pioneer defended. *New York Review of Books* 42 (20): 54–58.

Phillips, J. 1860. *Life on Earth: Its Origin and Succession.* London: Macmillan.

Pierce, J. 1823. A memoir on the Catskill Mountains, with notices of their mineralogy and zoology. *Silliman Journal* 6:86–97.

Pittenger, M. 1993. *American Socialists and Evolutionary Thought, 1870–1920.* Madison, Wisc.: University of Wisconsin Press.

Popper, K. R. 1945. *The Open Society and Its Enemies.* London: George Routledge and Sons.

———. 1959. *The Logic of Scientific Discovery.* London: Hutchinson.

———. 1972. *Objective Knowledge.* Oxford: Oxford University Press.

———. 1974. Intellectual autobiography. In *The Philosophy of Karl Popper,* ed. P. A. Schilpp. LaSalle, Ill.: Open Court.

Powell, B. 1855. *Essays on the Spirit of the Inductive Philosophy.* London: Longman, Brown, Green, and Longmans.

Presciuttini, S., L. Bertario, P. Sala, C. Rossetti, and R. C. Lewontin. 1993. Correlation between relatives for colorectal-cancer mortality in familial adenomatous polyposis. *Annals of Human Genetics* 57: 105–15.

Provine, W. B. 1971. *The Origins of Theoretical Population Genetics.* Chicago: University of Chicago Press.

———. 1986. *Sewall Wright and Evolutionary Biology.* Chicago: University of Chicago Press.

Putnam, H. 1981. *Reason, Truth, and History.* Cambridge: Cambridge University Press.

Rainger, R. 1991. *An Agenda for Antiquity: Henry Fairfield Osborn and Vertebrate Paleontology at the American Museum of Natural History, 1890–1935.* Tuscaloosa, Ala.: University of Alabama Press.

Rampino, M. R., and R. B. Stothers. 1984. Terrestrial mass extinctions, cometary impacts and the sun's motion perpendicular to the galactic plane. *Nature* 308: 709–12.

Raup, D. M. 1986. *The Nemesis Affair: A Story of the Death of Dinosaurs and the Ways of Biology.* New York: Norton.

Raup, D. M., and J. J. Sepkoski Jr. 1984. Periodicity of extinctions in the geologic past. *Proceedings of the National Academy of Science U.S.A.* 81: 801–5.

———. 1986. Periodic extinctions of families and genera. *Science* 231: 833–36.

———. 1988. Testing for periodicity of extinction. *Science* 241: 94–96.

Richards, R. J. 1992. *The Meaning of Evolution: The Morphological Construction and Ideological Reconstruction of Darwin's Theory.* Chicago: University of Chicago Press.

Richmond, M. 1988. Darwin's study of the Cirripedia. In *The Correspondence of Charles Darwin,* vol. 4. Cambridge: Cambridge University Press.

Ritvo, H. 1987. *The Animal Estate: The English and Other Creatures in the Victorian Age.* Cambridge, Mass.: Harvard University Press.

Robbins, B., and A. Ross. 1996. Letter to editor. *New York Times,* May 23.

Ruse, M. 1975. Darwin's debt to philosophy: an examination of the influence of the philosophical ideas of John F. W. Herschel and William Whewell on the development of Charles Darwin's theory of evolution. *Studies in History and Philosophy of Science* 6: 159–81.

———. 1979. *The Darwinian Revolution: Science Red in Tooth and Claw.* Chicago: University of Chicago Press.

———. 1980. Charles Darwin and group selection. *Annals of Science* 37: 615–30.

———. 1982. *Darwinism Defended: A Guide to the Evolution Controversies.* Reading, Mass.: Addison-Wesley.

———. 1986. *Taking Darwin Seriously: A Naturalistic Approach to Philosophy.* Oxford: Blackwell.

———. 1988. *But Is It Science? The Philosophical Question in the Creation/Evolution Controversy.* Buffalo, N.Y.: Prometheus.

———. 1989. *The Darwinian Paradigm: Essays on Its History, Philosophy and Religious Implications.* London: Routledge.

———. 1996. *Monad to Man: The Concept of Progress in Evolutionary Biology.* Cambridge, Mass.: Harvard University Press.

Ruse, M., and E. O. Wilson. 1986. Moral philosophy as applied science. *Philosophy* 61: 173–92.

Russell, E. S. 1916. *Form and Function: A Contribution to the History of Animal Morphology.* London: John Murray.

Russett, C. E. 1966. *The Concept of Equilibrium in American Social Thought.* New Haven, Conn.: Yale University Press.

———. 1976. *Darwin in America: The Intellectual Response. 1865–1912.* San Francisco, Calif.: Freeman.

———. 1989. *Sexual Science: The Victorian Construction of Womanhood.* Cambridge, Mass.: Harvard University Press.

Scheffler, I. 1967. *Science and Subjectivity.* Indianapolis, Ind.: Bobbs-Merrill.

Schiff, M., and R. C. Lewontin. 1986. *Education and Class: The Irrelevance of IQ Studies.* Oxford: Oxford University Press.

Schofield, R. E. 1963. *The Lunar Society of Birmingham: A Social History of Provincial Science and Industry in the Eighteenth Century.* Oxford: Clarendon Press.

Schwartz, R. D., and P. B. James. 1984. Periodic mass extinctions and the sun's oscillation about the galactic plane. *Nature* 308: 712–13.

Segerstrale, U. 1986. Colleagues in conflict: an in vitro analysis of the sociobiology debate. *Biology and Philosophy* 1: 53–88.

Selzer, J., ed. 1993. *Understanding Scientific Prose.* Madison, Wisc.: University of Wisconsin Press.

Sepkoski Jr, J. J. 1976. Species diversity in the Phanerozoic—species-area effects. *Paleobiology* 2: 298–303.

———. 1978. A kinetic model of Phanerozoic taxonomic diversity. I. Analysis of marine orders. *Paleobiology* 4: 223–51.

———. 1979. A kinetic model of Phanerozoic taxonomic diversity. II. Early Paleozoic families and multiple equilibria. *Paleobiology* 5: 222–52.

———. 1984. A kinetic model of Phanerozoic taxonomic diversity. III. Post-Paleozoic families and mass extinctions. *Paleobiology* 10: 246–67.

———. 1986a. Global events and the question of periodicity. In *Global Bio-Events: A Critical Approach*, ed. O. H. Walliser. Berlin: Springer-Verlag.

———. 1986b. Phanerozoic overview of mass extinctions. In *Patterns and Processes in the History of Life*, ed. D. M. Raup and D. Jablonski. Berlin: Springer-Verlag.

———. 1991. Diversity in the Phanevozoic oceans: a partisan review. In *The Unity of Evolutionary Biology. Proceedings of ICSEB IV*, ed. E. C. Dudley. Portland, Oregon: Dioscorides Press.

———. 1994. What I did with my research career: or how research on biodiversity yielded data on extinction. In *The Mass Extinction Debates: How Science Works in a Crisis*, ed. W. Glen. Stanford, Calif.: Stanford University Press.

Sepkoski Jr, J. J., and D. M. Raup. 1986a. Periodicity in marine extinction events. In *Dynamics of Extinction*, ed. D. Elliott. New York: John Wiley and Sons.

———. 1986b. Was there 26-Myr periodicity of extinctions? *Nature* 321: 533.

Sepkoski Jr, J. J., and M. A. Rex. 1974. Distribution of freshwater mussels: Coastal rivers as biogeographic islands. *Systematic Zoology* 22: 165–88.

Shapere, D. 1964. Review of *The Structure of Scientific Revolutions*. *Philosophical Review* 73: 383–94.

Shapin, S. 1982. History of science and its social reconstructions. *History of Science* 20: 157–211.

Simberloff, D. S., and E. O. Wilson. 1969. Experimental zoogeography of islands: the colonization of empty islands. *Ecology* 50: 278–96.

Simpson, G. G. 1944. *Tempo and Mode in Evolution*. New York: Columbia University Press.

———. 1949. *The Meaning of Evolution*. New Haven, Conn.: Yale University Press.

———. 1953. *The Major Features of Evolution*. New York: Columbia University Press.

Slobodkin, L. B. 1988. Review of *An Urchin in the Storm*. *American Scientist* 76: 503–4.

Smith, A. [1776] 1937. *The Wealth of Nations*. New York: Modern Library.

Smocovitis, V. B. 1992. Unifying biology: the evolutionary synthesis and evolutionary biology. *Journal of the History of Biology* 25: 1–66.

————. 1996. *Unifying Biology: the Evolutionary Synthesis and Evolutionary Biology.* Princeton, N.J.: Princeton University Press.

Sokal, A. D. 1996a. Transgressing the boundaries—toward a transformative hermeneutics of quantum gravity. *Social Text* 46–47: 217–52.

————. 1996b. A physicist experiments with cultural studies. *Lingua Franca,* May/June: 62–64.

Spencer, H. 1857. Progress: its law and cause. *Westminster Review* 67: 244–67.

————. 1862. *First Principles.* London: Williams and Norgate.

————. 1864. *Principles of Biology.* 2 vols. London: Williams and Norgate.

————. 1892. *The Principles of Ethics.* 2 vols. London: Williams and Norgate.

Stachel, J. 1995. Albert Einstein and Mileva Marie: a collaboration that failed to develop. In *Creative Couples in Science,* ed. H. Pycior, N. Slack, and P. Abir-am. New Brunswick, N.J.: Rutgers University Press.

Stebbins, G. L. 1950. *Variation and Evolution in Plants.* New York: Columbia University Press.

————. 1969. *The Basis of Progressive Evolution.* Chapel Hill, N.C.: University of North Carolina Press.

Stebbins, G. L., and F. J. Ayala. 1981. Is a new evolutionary synthesis necessary? *Science* 213: 967–71.

Stigler, S. M., and M. J. Wagner. 1987. A substantial bias in nonparametric tests for periodicity in geophysical data. *Science* 238, 940–45.

————. 1988. Reply to Raup and Sepkoski, "Testing for periodicity of extinction." *Science* 241: 96–99.

Sulloway, F. 1979. *Freud: Biologist of the Mind.* New York: Basic Books.

Teilhard de Chardin, P. 1955. *Le Phénomène humain.* Paris: Editions de Seuil.

Thomson, W. (Lord Kelvin). 1869. Of geological dynamics. *Popular Lectures* 2: 73–131.

————. 1884. *Notes of Lectures on Molecular Dynamics and the Wave Theory of Light.* Baltimore, Md.: Johns Hopkins Press.

Todes, D. P. 1989. *Darwin without Malthus: The Struggle for Existence in Russian Evolutionary Thought.* New York: Oxford University Press.

Trivers, R. L. 1972. Parental investment and sexual selection. In *Sexual Selection and the Descent of Man,* ed. B. Campbell. Chicago: Aldine-Atherton.

von Baer, K. E. [1828–37] 1853. Über Entwicklungsgeschichte der Thiere (Fragments related to Philosophical Zoology: selected from the works of K. E. von Baer). In *Scientific Memoirs,* ed. and trans. A. Henfry and T. H. Huxley. London: Taylor and Francis.

Vorzimmer, P. J. 1970. *Charles Darwin: The Years of Controversy.* Philadelphia: Temple University Press.

Wallace, A. R. 1870. *Contributions to the Theory of Natural Selection: A Series of Essays.* London: Macmillan.

————. 1905. *My Life: A Record of Events and Opinions.* 2 vols. London: Chapman and Hall.

Waters, C. K., and A. Van Helden, eds. 1992. *Julian Huxley: Biologist and Statesman of Science.* Houston, Tex.: Rice University Press.

Weinberg, S. 1996. Sokal's hoax. *New York Review of Books* 43 (13): 11–15.

Westfall, R. S. 1973. Newton and the fudge factor. *Science* 179: 751–58.

———. 1980. *Never at Rest: A Biography of Isaac Newton.* Cambridge: University of Cambridge Press.

Whewell, W. 1837. *The History of the Inductive Sciences.* 3 vols. London: Parker.

———. 1840. *The Philosophy of the Inductive Sciences.* 2 vols. London: Parker.

Whitmire, D. P., and A. A. Jackson. 1984. Are periodic mass extinctions driven by a distant solar companion? *Nature* 308: 713–15.

Wilson, E. O. 1959. Adaptive shift and dispersal in a tropical ant fauna. *Evolution* 13: 122–44.

———. 1961. The nature of the taxon cycle in the Melanesian ant fauna. *American Naturalist* 95: 169–93.

———. 1971. *The Insect Societies.* Cambridge, Mass.: Belknap Press of Harvard University Press

———. 1975. *Sociobiology: The New Synthesis.* Cambridge, Mass.: Harvard University Press.

———. 1978. *On Human Nature.* Cambridge, Mass.: Harvard University Press.

———. 1980a. Caste and division of labor in leaf cutter ants (Hymenoptera: Formicidae: *Atta*). I. The overall pattern in *Atta sexdens. Behavioral Ecology and Sociobiology* 7: 143–56.

———. 1980b. Caste and division of labor in leaf cutter ants (Hymenoptera: Formicidae: *Atta*). II. The ergonomic optimization of leaf cutting. *Behavioral Ecology and Sociobiology* 7: 157–65.

———. 1983. Caste and division of labor in leaf cutter ants (Hymenoptera: Formicidae: *Atta*). III. Ergonomic resiliency in foraging by *Atta cephalotes. Behavioral Ecology and Sociobiology* 14: 47–54.

———. 1984a. The relation between caste ratios and divisions of labor in the ant genus *Pheidole* (Hymenoptera: Formicidae). *Behavioral Ecology and Sociobiology* 16: 89–98.

———. 1984b. *Biophilia.* Cambridge, Mass.: Harvard University Press.

———. 1992. *The Diversity of Life.* Cambridge, Mass.: Harvard University Press.

———. 1994. *Naturalist.* Washington, D.C.: Island Books/Shearwater Books.

———. 1998. *Consilience: The Unity of Knowledge.* New York: Knopf.

Wilson, E. O., and D. S. Simberloff. 1969. Experimental zoogeography of islands: Defaunation and monitoring techniques. *Ecology* 50: 267–78.

Wright, R. 1987. *Three Scientists and Their Gods.* New York: Times Books.

Wright, S. 1931. Evolution in Mendelian populations. *Genetics* 16: 97–159.

———. 1932. The roles of mutation, inbreeding, crossbreeding and selection in evolution. *Proceedings of the Sixth International Conference of Genetics* 1:356–66.

Wright, S., and T. Dobzhansky. [1946] 1981. Experimental reproduction of some of the changes caused by natural selection in certain populations of *Drosophila pseudoobscura*. In *Dobzhansky's Genetics of Natural Population I–XLIII*, ed. R. C. Lewontin, J. A. Moore, W. B. Provine, and B. Wallace. New York: Columbia University Press.

Young, R. M. 1985. *Darwin's Metaphor: Nature's Place in Victorian Culture*. Cambridge, Mass.: Cambridge University Press.

Glossary

A PRIORI: a claim, as in logic and pure mathematics, the truth of which is not dependent on experiment or sense experience

ADAPTATION: any feature of an organism that aids survival and reproduction in a given environment

ALLELE: one of a number of versions of a gene that can occupy the same place (locus) on a chromosome

ALLELOMORPH: (obs.) allele

ALLOMETRY: study of the relative growth of parts of an organism in comparison with other parts

ALTRUISM: help given by one organism to another, at the immediate expense of the giver but for the giver's long-term reproductive advantages

AMINO ACIDS: complex organic molecules that are the building blocks of proteins

ANGIOSPERM: a plant whose seeds are enclosed in an ovary; loosely, a flowering plant

ARCHAEOCYATHID: an extinct form of sponge

ARMS RACE: a metaphor for competition between two species, as a result of which each species continuously adapts to changes in the other

ARTIFICIAL SELECTION: the methods used by animal and plant breeders to alter a species, through picking and breeding favored variants

AUTOCATALYTIC: becoming more powerful as a result of feedback mechanisms

BACONIAN INDUCTIVISM: the collecting of facts in order to draw generalizations

BALANCE HYPOTHESIS: the claim that natural selection holds many different alleles in balance or equilibrium within a population

BALANCE OF NATURE: the pre-Darwinian notion that God has so designed organisms that their numbers remain in equilibrium

BALANCED SUPERIOR HETEROZYGOTE FITNESS: the claim that selection keeps different alleles in equilibrium within a population, because the heterozygote is fitter than either homozygote

BAUPLAN: the basic plan or archetype of an organism

BIODIVERSITY: the numbers and kinds of organisms and their mutual relationships

BIOGENETIC LAW: the claim that ontogeny (the development of individuals) recapitulates phylogeny (the evolution of a particular line)

BIOGEOGRAPHY: the study of the spatial distributions of organisms

BIOLOGICAL DETERMINISM: the claim that the traits of organisms are a direct function of their genes, not subject to environmental influences

BIOLOGICAL PROGRESS: the claim that evolution shows a progressive rise from the simple to the complex, the latter being best represented by humankind

BIOMETRICIANS: Darwinian selectionists at the beginning of this century who were critical of Mendelism, believing rather that variation is blending and continuous

BURGESS SHALE: an outcrop in the Canadian Rockies containing fossilized soft-bodied animals of the early Cambrian

CAMBRIAN EXPLOSION: the great increase in biodiversity that began about 570 million years ago

CATASTROPHISM: the belief that earth's history has been marked by major upheavals that led to mass extinctions

CHROMOSOME: the threadlike entities within the center of the cell that carry the genes

CLADE: all of the organisms descended from one particular taxon

CLASSICAL HYPOTHESIS: the claim that little genetic variation exists within a population, thanks to the purifying effect of selection

CLASSICAL THEORY OF THE GENE: the theory that integrates Mendelian genetics with cytological discoveries by locating the gene on the chromosome

COHERENCE: the epistemic value of having the parts of a theory hang together consistently

COHERENCE THEORY OF TRUTH: the claim that truth consists in getting one's beliefs to mesh together, irrespective of whether or not there is a real world to which they correspond

CONSILIENCE OF INDUCTIONS: William Whewell's term for the epistemic value of bringing many bodies of knowledge together under one unifying idea or system

CONSISTENCY: the epistemic value of not being in conflict with other ideas or systems

CONSTRAINT: any aspect of organic physiology or morphology that prevents natural selection from working in the simplest and most advantageous direction

CONTEXT OF DISCOVERY: the processes whereby a scientific claim is made or developed

CONTEXT OF JUSTIFICATION: the processes whereby a scientific claim is validated

CORRESPONDENCE THEORY OF TRUTH: the claim that knowledge corresponds truly to a mind-independent reality

CREATION SCIENCE: a system claiming that scientific evidence supports the story of creation as told in the early chapters of Genesis

CULTURAL VALUES: interests or norms reflecting societal standards

DARWINISM: the claim that natural selection is the overwhelmingly significant causal factor in evolution

DEISM: the belief that God is an Unmoved Mover who works only through unbroken law, without using miracles

DIPLOID: having a paired set of chromosomes

DIVISION OF LABOR: the process of breaking down a function into more specialized tasks, with the intention of performing the function more efficiently

DNA (DEOXYRIBONUCLEIC ACID): the macromolecule that transmits genetic information

DROSOPHILA: the fruit fly, a favorite organism for study by geneticists

DYNAMIC EQUILIBRIUM: a balance between opposing forces that is constantly in motion, often upward

ECOLOGY: the scientific study of the interrelations of organisms in nature

EMBRYOLOGY: the scientific study of the development of organisms during the embryonic stage

EMPIRICISM: the philosophy that knowledge must begin with experience, particularly of the senses

ENCEPHALIZATION QUOTIENT (EQ): a measure of intelligence across species, based on comparison of brain and body size

EPISTEMIC VALUES: those norms or rules that supposedly lead to objective knowledge

EPISTEMOLOGY: the branch of philosophy that studies the nature and origin of knowledge

EQUILIBRIUM: a state of balance between opposing forces

EUGENICS: the claim that the way to improve humankind is through selective breeding

EVOLUTION: the change in groups of organisms over time, so that descendants differ from their ancestors

EVOLUTIONARY EPISTEMOLOGY: any theory of knowledge using or modeled on evolutionary thinking

EVOLUTIONARILY STABLE STRATEGY (ESS): a genetically programmed strategy taken by members of a population such that no other strategy can dislodge it

EXAPTATION: an organic feature without adaptive function

EXTERNALISM: the historical approach that attempts to understand scientific change in terms of social factors outside of science itself

FALSIFIABILITY: the possibility of a scientific system's being refuted by experience

FERTILITY: the epistemic value of stimulating new ideas or directions of research

FITNESS: relative ability of an organism to get its genes into the gene pool of the next generation

FOUNDER PRINCIPLE: the claim that new species form when a small group of organisms with atypical genotypes become isolated from the larger population and interbreed

FUNCTION: the end toward which an organic adaptation is directed

GEL ELECTROPHORESIS: a technique for detecting variations in molecular genes by tracking their movement through a gel under the influence of an electric field

GENE: the unit of heredity, which today is known to consist of a sequence of base pairs in a molecule of DNA

GENE POOL: the collective genes of a population or species

GENETIC DRIFT: the claim that in small populations accidents of mating can outweigh any effects of natural selection

GENOME: the genetic material of an organism

GENOTYPE. an organism's genetic information, as distinguished from its physical appearance (phenotype)

GRADUALISM: the belief that organic change occurs gradually

GROUP SELECTION: the claim that natural selection operates through adaptations which benefit the group, at the expense of individuals

HAPLOID: the half set of chromosomes received from father or mother

HETEROZYGOTE: an organism with two different alleles at some locus, as opposed to two identical alleles

HIERARCHY: the claim that existence occurs on many levels and that each level has unique characteristics

HOLISM: the claim that new properties emerge in "wholes" that cannot be found among their parts, and that entities, therefore, are more than the sum of their parts

HOMEOSTASIS: the state of balance or equilibrium in which different forces are held within an individual or population

HOMOLOGY: a similarity or isomorphism between organisms of different species, due to their common ancestry

HOMOZYGOTE: an organism with two identical alleles at some locus, as opposed to two different alleles

HYMENOPTERA: the ants, bees, and wasps

HYPOTHETICO-DEDUCTIVE SYSTEM: a system where empirical laws (like Kepler's laws of planetary motion) can be deduced from high-level hypotheses (like Newton's laws of motion and of gravitational attraction)

IDEALISM: the belief that the world in some sense is dependent on the perceiving subject

INCLUSIVE FITNESS: the reproductive fitness of an individual, together with the reproductive fitness of relatives who share some of its genes

INCOMMENSURABLE: having no way to translate from one language or set of ideas to another

INDIVIDUAL SELECTION: the claim that natural selection operates through favoring the adaptations of individual organisms, as opposed to groups

INTERNAL REALISM: a form of idealism making reality dependent on the perceiving subject

INTERNALISM: the historical approach that attempts to understand scientific change solely in terms of ideas within science itself

IRIDIUM: a rare chemical element found in platinum ores

ISOLATING MECHANISMS: physiological, morphological, or behavioral barriers to reproduction between members of different species

JUST SO STORIES: improbable fantasies spun to give an adaptive explanation of puzzling organic features

KIN SELECTION: the claim that natural selection operates through the benefits an organism confers on its close relatives which increase their reproductive fitness and hence the organism's own inclusive fitness

KINETIC: pertaining to movement or motion

LAMARCKISM: the claim that evolution occurs because offspring inherit characteristics that their parents acquired in response to changes in their environment

LOCUS: a particular area or point on a chromosome occupied by different forms (alleles) of a gene

LOGICAL POSITIVISM: a twentieth-century philosophy that tried to reduce all knowledge either to verifiable sense experience or to the relations and meanings of words

MACROEVOLUTION: evolution occurring above the species level, over large periods of time

MACRO-MUTATION: mutation which causes major changes; a saltation

MENDELISM: a theory of genetics, based on Mendel's laws, which stressed that the units of inheritance do not change or blend when passed along to offspring (except in the case of drastic mutations) and that offspring can carry genes for traits which may not be expressed until a later generation

METAPHYSICAL REALISM: the claim that a real world exists independent of our experience; also known as realism

METAPHYSICS: that branch of philosophy which deals with the ultimate nature and origin of things

METAVALUE: a value about the nature of science rather than a value within science itself

METAZOAN: a multicellular animal

MICROEVOLUTION: evolution occurring at or below the species level, over short periods of time

MOLECULAR BIOLOGY: the branch of biology which focuses on organic processes at the level of molecules, as opposed to cells, organs, whole organisms, populations, species, communities, and so on

MOLECULAR DRIFT: the claim that selection has no significant effect at the molecular level, and therefore variation occurs randomly

MORPHOLOGY: the scientific study of organic form

MUTATION: a change in the genetic makeup of an organism, traditionally identified by changes in its physical or behavioral characteristics; today, a spontaneous or induced change in the DNA sequence of a gene, identified through DNA analysis

NATURAL SELECTION: the key Darwinian mechanism for evolutionary change, claiming that a small percentage of organisms in each generation survive and reproduce owing to characteristics which other members of the population do not possess; these adaptive characteristics are passed along to offspring

NATURAL THEOLOGY: the attempt to understand the nature of God through reason or experience rather than through faith and revelation

NATURALISM: the attempt to understand by reference to unbroken law

NATURALISTIC FALLACY: the attempt to derive claims about morality from facts about the world

NATURPHILOSOPHIE: an early nineteenth-century German philosophy of morphology, stressing homology (similarities between the parts of organisms of different species) over adaptation (function)

NEO-DARWINISM: the synthesis of Darwinism with Mendelism

NEUTRAL EVOLUTION: the accumulation of heritable mutations that do not promote or reduce the fitness

NONEPISTEMIC VALUES: cultural values (religious, racial, sexual, and others)

NORMAL SCIENCE: science that is done within a paradigm

NUCLEIC ACID: a chainlike macromolecule found in cells, either DNA (deoxyribonucleic acid), which carries the information of heredity, or RNA (ribonucleic acid), which reads the information from the DNA

OPTIMIZATION: the action of natural selection to produce the most efficient of all possible adaptations

ORTHOGENESIS: the claim that life develops according to a predetermined momentum, and is not subject to the influence of external factors

OVIPOSITION: egg laying

PALEOANTHROPOLOGY: the scientific study of the evolution of humankind, with a focus on the fossil evidence

PALEONTOLOGY: the scientific study of the fossil record

PANGENESIS: Charles Darwin's theory of heredity which supposes that particles from all parts of the body are carried to the reproductive organs, there forming the sex cells

PANGLOSSIANISM: the claim that every aspect of organisms has some adaptive function

PANPSYCHIC MONISM: the philosophy that everything is a manifestation of a universal mind

PARADIGM: a body of work or ideas with its own internal truth, providing the basis for everyday scientific practice (normal science), and replaced only by a drastic rupture or revolution

PARENTAL INVESTMENT: the efforts expended by parents on their offspring

PARENT-OFFSPRING CONFLICT: competition between parents and offspring over the allocation of resources

PHENOTYPE: the physical and behavioral characteristics of an organism

PHRENOLOGY: the pseudo-science which holds that character can be read from the bumps on the skull

PHYLOGENY: a particular path of evolution (for instance, that from reptiles to birds)

PHYSIOLOGY: the scientific study of the workings of living bodies and their parts

PLATONISM: the philosophy of the Greek philosopher Plato, centering on ideals or Forms that exist in a nonphysical world of ultimate reality, accessible only through trained rational intuition

POLYMORPHISM: physical or behavioral variation within a group

POPULAR SCIENCE: science writing aimed deliberately at the general public

POPULATION GENETICS: the extension of Mendelian genetics to groups, showing how balance and chance can occur thanks to such causal factors as selection and mutation

POSITIVISM: an empiricist philosophy tending to be hostile to metaphysical claims about such things as ultimate reality

PREDICTIVE ACCURACY: the epistemic value of making accurate predictions

PROGRESS: see social progress; biological progress

PROVIDENCE: God's actions in and intentions for the world, especially for humankind

PSEUDO-SCIENCE: a body of belief claiming to be genuine science but driven by cultural beliefs to the detriment of epistemic standards

PSYCHOLOGISM: the claim that the context of discovery is pertinent to the context of justification

PUNCTUATED EQUILIBRIA: the claim that evolution consists of stability (stasis) interrupted by rapid major change

REALISM: the claim that a real world exists independent of our experience; often now called metaphysical realism

RECIPROCAL ALTRUISM: help given by organisms to one another, with the expectation that the help will be returned

REDUCTIONISM: the claim that aspects which organisms exhibit at a higher level can be explained fully in terms of processes occurring at lower levels; the opposite of holism

REVEALED RELIGION: belief based on revelation (such as the Bible), generally involving faith

REVOLUTIONARY SCIENCE: science which breaks from one paradigm and switches to another; the opposite of normal science

SALTATIONISM: the claim that evolutionary change occurs through sudden jumps or macro-mutations

SEWALL WRIGHT EFFECT: genetic drift

SEXUAL SELECTION: the claim that natural selection operates through competition over mates

SHIFTING BALANCE THEORY: the claim that a balance exists between those forces leading to genetic similarity or homogeneity and those leading to genetic diversity or heterogeneity

SIMPLICITY: the epistemic value of being able to explain with few elements in an elegant fashion

SOCIAL CONSTRUCTIVISM: a form of idealism claiming that scientific ideas are epiphenomena of social or cultural ideas or movements

SOCIAL DARWINISM: any kind of social or political philosophy based on evolutionary principles but most often used (negatively) of laissez-faire doctrines associated with Herbert Spencer and his followers

SOCIAL PROGRESS: the philosophy of history that sees a gradual improvement in society or culture

SOCIOBIOLOGY: the scientific study of the biological basis of social behavior, with special emphasis on reproductive behavior

SPECIATION: the process whereby new species form

SPECIES: a group or population of interbreeding organisms reproductively isolated from all others

SPECIES SELECTION: the claim that natural selection operates through the differential success of species, which leads to the extinction of some and the origin of new ones

STABILIZING SELECTION: the claim that natural selection holds many different forces in balance or equilibrium

STRUGGLE FOR EXISTENCE: the effort expended by organisms, usually in competition with others, to survive and reproduce

SUPRAORGANISM: a group of organisms so integrated that selection acts on them as one individual

SURVIVAL OF THE FITTEST: a term coined by Herbert Spencer and later used by Charles Darwin to refer to natural selection

SYNTHETIC THEORY OF EVOLUTION: the integration of Darwinian selection with Mendelian genetics; neo-Darwinism

SYSTEMATICS: the scientific theory behind the classification of organisms

TAXON (PLURAL TAXA): a group of organisms which share a common ancestry and which are sufficiently distinct to merit a formal name

TAXONOMY: the practice of classifying organisms

TELEOLOGY: the study of end-directed processes in nature, stressing the interconnectedness of all of reality

TETRAPODS: four-limbed animals

THEISM: the belief that God intervenes in His Creation

TRANSCENDENTALISM: an idealistic philosophy best represented by *Naturphilosophie*

TROPHIC: having to do with the processes of nutrition

ULTRA-ADAPTATIONISM: the belief that everything has an adaptive function; Panglossianism

UNIFORMITARIANISM: the claim that all events of the geological past can be explained in terms of processes analogous to those operating today

UNIFYING POWER: the epistemic value of being able to integrate diverse items into one coherent system

VITALISM: the claim that organisms are driven by nonmaterial life forces

WHIGGISHNESS: the inclination to interpret events in the past in terms of their contribution to progress and to the superiority of the present

Index

Academy of Sciences (French), 47
Achinstein, P., 15
adaptation: early views on, 41–42, 56–57, 69, 91, 101; present views on, 138, 154, 166–167, 182, 228, 231, 232. *See also* nonadaptive; selection: natural
adaptationism, 127, 141, 149, 159, 164, 168, 202–203
adaptive landscape, 85–87, 101–102, 109, 115, 118–119, 239. *See also* Wright, S.
Agassiz, L., 73
aggression, 178, 198–199, 202
agnosticism, 68–69
Alabama, University of, 188, 191
Albon, S. D., 201
alcohol tolerance, 107
allele, 82, 86, 156
allelomorph. *See* allele
Allen, G., 81
allometry. *See* relative growth
altruism, 178–179, 189, 248
Alvarez, L., 224–225
America (U. S.), 88, 97, 100–101, 108, 159, 240; education, 144; South, 172, 187–191, 239; way of life, 79, 165, 232
American Civil Liberties Union, 135
American Museum of Natural History, 144
amino acids, 156
analogy, 61, 131, 253. *See also* metaphor
anarchism, 247

angiosperms. *See* flowering plants
ants, 60, 172, 179–182, 187, 192, 242–245; leaf cutter, 173, 179–182. *See also* social insects
Archaeocyathids, 230
archetype, 138, 141, 144, 159–160
argument from design, 41, 52, 79, 130, 144. *See also* adaptation
Aristotle, 241
Arkansas, 135
arms race, 91, 93–94, 131–132, 239
Arnold, M., 97
Arnold, T., 96
artificial selection. *See* selection: artificial
astrophysics, 224–227, 230
atheism, 130–132
Atomic Energy Commission, 111
autocatalytic model, 178, 187, 190
Ayala, F. J., 107, 138, 169

Babbage, C., 44
baboon, 202
Baker, J. R., 98
balance hypothesis, 109–110, 156–157, 163–164, 166
balance of nature, 79, 154, 173, 183, 189–190, 251. *See also* equilibrium
balanced superior heterozygote fitness, 105–106, 109–110, 157, 191
Barbour, I., 44
Barker, S., 15

nonadaptive, 136, 138–142, 203, 220
normal science. *See* Kuhn, T. S.: normal
  science
Notre Dame University, 231
nuclear tests, 110–111
nuclei, 82
Numbers, R., 73
Nyhart, L., 77

objectivity. *See* science
observation, 196
Open Society and Its Enemies. See Popper,
  K. R.
optimality models. *See* optimize
optimize, 180–182, 198, 203, 245
order (category), 217, 229
Origin of Species. See Darwin, C.
orthogenesis, 73
Osborn, H. F., 90, 168
Outram, D., 75
oviposition, 197–198
Owen, R., 246
Oxford University, 88, 97, 100, 129, 132

paleoanthropology, 123
Paleobiology, 146, 151
paleontology, 90–91, 108, 136, 142–144, 147,
  214–235; Darwin's position, 60, 66–67.
  See also fossils; geology
Paleozoic, 220–222, 227, 233
Paley, W., *Natural Theology,* 41, 68
pangenesis, 64
Panglossianism, 141, 159, 180. *See also*
  adaptationism
panpsychic monism, 87
paradigm. *See* Kuhn, T. S.: paradigm
paradox: biodiversity, 183; social behavior,
  177–178, 200
parent-offspring conflict, 200–202, 205,
  210–211
parental investment, 126. *See also*
  parent-offspring conflict
Parker, G., 124, 194–213, 235, 237, 239, 248,
  250, 252
Pasteur, L., 7–8, 10
Patterson, C., 226
Pearson, K., 87
Pearson, R. G., 203
Peirce, C. S., 72
pendentive, 139

periodicity (of extinction), 223–227, 229–230
Permian, 222, 223
Perutz, M., 10
Phanerozoic, 216, 219–220, 223, 229, 232
phenotype, 82, 155
Phillips, J., 65
philosophy, 9, 85, 171, 182, 184, 210, 253–255;
  pejorative sense of, 120
phrenology, 27
phylogeny. *See* evolution: path
physics, 65
Pierce, J., 62
Pittenger, M., 79
Plato, 12
Pleistocene, 227
Pliocene, 227
poetry, 38, 42, 51
Popper, K. R., 13–18, 23, 28–29, 33,
  249–250; knowledge, 35, 236; *Logic of
  Scientific Discovery,* 14, 18, 25, 254–255
popular science, 117, 119, 121, 124–152, 167,
  192; Darwin's work as, 77, 83
population genetics, 83–88, 91–92, 156, 158,
  162–163
positive heuristic, 241. *See also* fertility
Powell, B., 44–45
Pre-Cambrian boundary, 219
predator-prey, 62
predictive accuracy, 32; modern, 162, 184,
  195, 198–199, 205–206, 229–230, 244;
  pre-modern, 43, 48, 62–63, 93, 107
predictive fertility. *See* fertility
Prescuittini, S., 170
Priestly, J., 45
principle of divergence, 58, 66, 242
Private Science of Louis Pasteur, 8, 10
problems. *See* science
*Proceedings of the National Academy of
  Sciences,* 224
professional science, 76, 117–119, 121, 136,
  168, 207, 254–255; achieved, 158, 235,
  236–237; not achieved, 47–49, 89, 96,
  98–99, 152; teamwork, 123–124. *See also*
  Gould, Stephen Jay: status as scientist;
  pseudo science
progress, 69–70, 87, 146, 164, 208, 239;
  biological, 69–70, 117, 119, 131–132, 145,
  221–222, 233; religion substitute, 45–46,
  78–80, 94–96, 98–99, 189–190; scientific,
  23, 208; with religion, 79, 108–109